SLEEP AND BREATHING IN INFANTS AND YOUNG CHILDREN

SLEEP AND BREATHING IN INFANTS AND YOUNG CHILDREN

IGOR A. KELMANSON

Nova Biomedical Books
New York

For permission to use material from this book please contact us:
Telephone 631-231-7269; Fax 631-231-8175
Web Site: http://www.novapublishers.com

Library of Congress Cataloging-in-Publication Data
Kelmanson, I. A. (Igor Aleksandrovich)
Sleep and breathing in infants and young children / Igor A. Kelmanson.
p. ; cm.
Includes bibliographical references and index.
ISBN 1-60021-279-4
1. Infants--Sleep. 2. Children--Sleep. 3. Sleep disorders in children. I. Title.
[DNLM: 1. Sleep Disorders. 2. Child. 3. Infant. 4. Respiratory System--physiopathology. 5. Sleep--physiology. 6. Sudden Infant Death. WM 188 K292s 2006]
RJ506.S55K45 2006
618.92'8498--dc22 2006017869

Published by Nova Science Publishers, Inc. ✣New York

Contents

Preface

This book addresses one of the major issues of modern pediatrics: relationship between sleep and breathing in infants. The content of this book is based on author's personal experience, as well as on up-to-date findings in the field. The necessity to present most important knowledge on sleep physiology, sleep structure in infants, mechanisms supporting respiration in the normal infants, functional changes in certain sleep stages, as well as on most characteristic forms of breathing disturbances in sleep in the infants was taken into consideration. Chronobiological aspects of functional respiratory characteristics have been tackled. Much attention has been paid to the quality of infant behavior associated with sleep disturbances and breathing disturbances in sleep, relationship between micro environmental characteristics and sleep impairments. Possible association between infant apnea and sudden infant death syndrome is considered; mechanisms of infant' snoring in sleep and its relation to the obstructive sleep apnea syndrome is addressed; sleep-related breathing disturbances occurring against a background of chronic broncho-pulmonary diseases are discussed. Possible approaches to clinical studies in infants with sleep-related breathing disturbances and to their treatment are considered.

The book is addressed to the broad audience including general pediatricians, pediatric pulmonologists, neurologists, ENT-specialists, intensive care specialists, somnologists, and clinical pharmacologists.

List of Abbreviations

ADHD	attention deficit/hyperactivity disorder
ALTE	apparent life threatening events
AP	apnea of prematurity
BA	bronchial asthma
BP	blood pressure
BPD	broncho-pulmonary dysplasia
CA	central apnea
CCHS	congenital central hypoventilation syndrome
CHVS	central hypoventilation syndrome
CI	confidence interval
CNS	central nervous system
CSF	cerebrospinal fluid
CT	computed tomography
DRG	dorsal respiratory group
ECG	electrocardiogram
EEG	electroencephalogram
EMG	electromyogram
EOG	electro-oculogram
GER	gastro-esophageal reflux
GERD	gastro-esophageal reflux disease
HD	Hirschsprung's disease
HR	heart rate
IA	infantile apnea
IUGR	intrauterine growth retardation
LBW	low birth weight
MA	mixed apnea
MCADD	medium chain acyl-coA dehyrogenase deficiency
MRI	magnetic resonance imaging
MSLT	multiple sleep latency test
NIPPV	noninvasive positive pressure ventilation
OA	obstructive apnea

OSAS	obstructive sleep apnea syndrome
PB	periodic breathing
PGO	ponto-geniculo-occipital
PSG	polysomnography
PV	pulmonary ventilation
RAR	rapidly adapting receptors
REM	rapid eye movement
SAR	slowly adapting receptors
SD	standard deviation
SEM	standard error of the mean
SIDS	sudden infant death syndrome
SWS	slow wave sleep
UAW	upper airway
UARS	upper airway resistance syndrome
UGI	upper gastrointestinal
UPPP	uvulo-palato-pharyngo-plastic
URTI	upper respiratory tract infection
VRG	ventral respiratory group

Introduction

The sources of modern clinical somnology, a part of clinical medicine that addresses sleep physiology and abnormalities, date back to the 1950-ies, when so-called paradox-sleep has been described. This led to the understanding of sleep heterogeneity and certain phenomena related to specific sleep stages. By the mead 1970-ies, electrophysiological methods of studies in sleep have been widely implemented, which has been defined as polysomnographic study. As a result, serious shifts in the mentality of the medical care providers have been achieved, and it became obvious that the evaluation of sleep structure and related physiological phenomena plays critical role from the standpoint of correct diagnosis and treatment of different functional disturbances, both presenting at the day time and in sleep. To-date, sleep studies are usually performed at the specialized centers and laboratories. Particularly, by the end of 2004, it was only in Germany that had approximately 300 officially registered and certified sleep laboratories, united by the German Sleep Society. Similar situation is characteristic of other industrialized countries. As a rule, the staff of these laboratories is composed of the clinical experts in sleep disturbances (sinologists), technicians, as well as consultants and sub-specialists, including ENT, cardiologists, neurologists, psychiatrists, paleontologists and others.

From the technical and methodological points of view, studies of sleep and sleep-related phenomena in the infants are much more difficult than in the adults. Specific and very important issue of pediatric somnology is possible association of infant sleep and certain forms of sleep-disordered breathing. Respiratory disorders are well acknowledged by general pediatricians among the leading causes of infant morbidity, mortality and chronic disability. However, they are not well aware of possible relationship between respiratory disorders in the infants with sleep and its certain stages. Meanwhile, infant sleep is the critical condition when respiratory disturbances may become most evident, and sometimes they can be only detected at sleep. Some forms of sleep disturbances and sleep disordered breathing may manifest at day time as the specific features of infant behavior, and the knowledge in that field is critical to make proper choice in diagnostic evaluation and treatment.

Micro-environmental factors may also modify infant's behavior, when he/she is either awake or asleep. It has been suggested that one of the extreme outcomes of sleep-disordered breathing may present as the sudden infant death syndrome, one of the mysterious and still poor understood problems of modern pediatrics. Furthermore, it has been found that such an "innocent" phenomenon as snoring may occur as early as the infancy, and it serve as the clinical marker of severe respiratory abnormality in sleep and may lead to numerous functional disturbances. So-called chronic broncho-pulmonary diseases, including asthma may sometimes manifest in infant sleep.

All the above was the reason of wide implementation of somnology in clinical pediatrics, and it was in the 1980-90-ies that well-equipped sleep laboratories have emerged worldwide. The example was one of the leading European sleep laboratories at the University Children's Hospital Queen Fabiola (Brussels, Belgium). The founder of this laboratory and its head for a long time was Professor André Kahn, world-renowned scientist. The author of this book was lucky to be personally acquainted with this outstanding person. The acquaintance first happened in Travemünde (Germany) in 1992 at the Congress of the European Society for the Study and Prevention of Infant Death (ESPID), now renamed to International Society for the Study and Prevention of Infant Death (ISPID). At that time, the Society just made its first steps, and Professor Kahn was one of the founders and the first President of the ESPID. The author of this book had a chance to work one year at the sleep laboratory headed by Professor Kahn. Scientific and personal contacts have been continued later on. Looking back at the scientific heritage of Professor Kahn, one can say without any exaggeration that there was not any hot issue of pediatric sleep medicine and related topic that has not been the subject of scientific concern of Professor Kahn. Many findings obtained during the fruitful years of our scientific collaboration are presented in this book. The author feels obliged to express the words of love and deepest gratitude to his late Friend and Master.

This book is aimed at giving basic knowledge on infant's sleep neurophysiology and on the mechanisms of respiratory control in sleep and in wake, as well as to shed the light on major functional changes related to certain sleep stages. Particular attention is paid to the factors predisposing to the acute breathing disturbances in sleep in the infants. Approaches to diagnosis and treatment of sleep-related respiratory disturbances in the infants are also considered.

The author relies upon interested attitude of medical audience and readily and gratefully will accept valuable and constructive criticism.

Part 1. Normal Sleep and Breathing in Infants and Young Children

Chapter I

Polysomnography as the Objective Method of Sleep Study

Sleep studies became widely available after invention of *polysomnography (PSG)*. Polysomnography refers to the collective process of monitoring and recording physiologic data during sleep. A polysomnogram consists of a simultaneous recording of multiple physiologic parameters related to sleep and wakefulness. The interaction of various organ systems during sleep and wakefulness is also evaluated. PSG is used to evaluate abnormalities of sleep and/or wakefulness and other physiologic disorders that have an impact on or are related to sleep and/or wakefulness. The variables monitored and recorded during polysomnography: global neural electroencephalographic activity (EEG) from electrodes placed on the patient's scalp; eye movements (electro-oculogram, or EOG) from electrodes placed near the outer canthus of each eye; submental electromyographic activity (EMG),rhythm electrocardiogram (ECG), respiratory effort, nasal and/or oral airflow via thermistor or pneumotachograph; oxygen saturation (SpO2) via pulse oximetry; body position via mercury switches or by direct observation; limb movements (arms and legs) via EMG; recordings of or vibration (frequency and/or volume) may be recorded; end-tidal CO2, transcutaneous CO2, esophageal pH, and some other parameters, if required. Information is gathered from all leads and fed into a computer and outputted as a series of waveform tracings which enable the technician to visualize the various waveforms, assign a score for the test, and assist in the diagnostic process.

Six EEG electrodes (labeled C_3, C_4, A_1, A_2 O_1, and O_2) and one ground electrode are placed around the cranium to record electrical activity across the brain. These leads are used to determine the stage of sleep the patient is in during any given period of the night (Sheldon HS et al., 1992). It is sufficient to have one EEG channel for appropriate sleep stage scoring, however many researchers prefer to use additional channels for more complete information. In particular, recordings from the occipital electrodes may be helpful in better identification of alpha-rhythm; coronar and parasaggital bipolar leads may help to identify of epileptic activity. Recordings from all EEG channels are performed when specific abnormal bio-electric activity should be looking for.

EOG is recorded with one electrode placed above and to the outside of the right eye, and another electrode is placed below and to the outside of the left eye. These leads record the movements of the eyes during sleep and serve to help determine sleep stages.

Three leads of EMG are placed on the chin (one in the front and center and the other two underneath and on the jawbone) and two are placed on the inside of each calf muscle 2-4cm apart. These leads serve to demonstrate muscle movement during sleep. This is helpful in documenting a wake period, an arousal, or just a spastic movement.

Two ECG electrodes are placed on the upper chest near the right and left arms. These record the heart rate and rhythm and serve to alert the technician to a possible emergency situation. They also demonstrate whether apneic desaturation leads to arrhythmias or not.

Airflow recording device (thermistor or thermocouple sensor) looks similar to a nasal cannula, and is secured just under the patient's nose. It senses the amount of air moving into and out of the airways and sends a signal to a physiological recorder. This tracing is used to determine the presence and extent of apneic episodes. Respiratory effort is usually recorded with piezo crystal effort sensor. Two bands, one placed around the chest under the breasts and one around the abdomen, serve to determine chest wall and abdominal movements during breathing. Each band is joined together by a piezo crystal transducer. The force of chest/abdominal expansion on the bands stretches the transducer and alters the signal to a physiological recorder. These leads, combined with the airflow sensor, are how apnea is demonstrated and categorized during the test.

If the sleep disorders center is equipped with video cameras in the patient rooms, the patient can be taped while sleeping. This allows the technician to review the tape at any time during the test and verify whether strange looking waveforms were caused by an actual arousal, a period of wake, or normal patient movement in bed.

By international standards, a polysomnogram must have a minimum of 4 neurophysiologic channels: one EEG channel to monitor sleep stage, two EOG channels to monitor both horizontal and vertical eye movements, one EMG channel to record atonia of rapid eye movement (REM) sleep.

Each sleep study is scored epoch by epoch both for stage of sleep and any abnormalities that can be seen. An epoch is a convenient time interval, usually equal to one page of record. Epoch durations should be 20 - 30 sec respectively depending on whether the recommended 15 or 10 mm/sec paper speeds are used. Automatic, computer-based systems often are employed in clinical and research settings. However, standard analysis still consists of tedious and time-consuming review and scoring of either paper tracings or recordings projected on a computer monitor.

In 1992, the Office of Technology Assessment of the Agency of Health Care Policy and Research recommended, in an evidence-based assessment, 2 tests as having been studied sufficiently. Both tests are performed in a sleep laboratory. These are overnight polysomnography, which is an overnight recording of the patient's sleep, and multiple sleep latency testing (MSLT) that records multiple naps throughout a day.

Standard sleep studies usually include both tests. Limitations usually stem from the fact that recording conditions may not reflect what happens during a regular night in the patient's home.

Although diagnosing a sleep problem on the basis of a recording over a single night is common practice, some authorities caution that more than 1 night of recording may be necessary, so the patient may become comfortable with unfamiliar surroundings and sleep more naturally. This effect is greatest on the first night in the sleep laboratory ("first night effect"). Sporadic events may be missed on a 1-night PSG. External factors that disturb the subject's sleep may be present in the home but absent from the controlled environment of the sleep laboratory. Patient preparation is important so that the patient sleeps naturally. High costs and long waiting lists have prompted the exploration of alternative methods of evaluation, including ambulatory monitoring with portable equipment, daytime PSG, simplified sleep studies with limited subsets of monitored parameters.

Overnight parameters (e.g., times of lights on/off, total time in bed, total sleep time) are collected. The standard EEG, EMG, and EOG recordings are evaluated, and the predominant stage of sleep then is assigned to the entire epoch. *Total time* and *relative proportion of the night spent in each of sleep stages* are calculated. *Sleep efficiency* is calculated as a proportion of time the patient was sleeping to the total time of PSG recording. *Latencies* to REM and slow-wave sleep (SWS) are reported. *Sleep latency* is determined as the time passed since lights off to the appearance of three consecutive 30-second epochs corresponding to sleep stage 1 or first 30-second epoch of any other sleep stage. *Latency to REM-sleep* is the time that passed from the beginning of sleep till the first 30-second epoch corresponding to REM-sleep. *Hypnogram,* known as a schematic presentation of transitions from one stage to another across the night, is also analyzed. Special neurophysiologic events (eg, epileptic events, intrusion of alpha into sleep, periodic activity of m. tibialis anterior) are reported. Respiratory activity is correlated with sleep stages. Other parameters such as body position, gastroesophageal reflux, bruxism, are recorded.

Chapter II

Basic Knowledge on Sleep Neurophysiology

The state of wakefulness is assured by the function of integrated system including at least 10 structures of brain. Sleep onset is observed when the wakefulness is suppressed by neuronal structures that synthesizes hypnogenic factors (Sakai K, 1988). It is impossible to identify exact neuronal structures of the brain responsible for the state of wakefulness. At the same time, certain cellular structures located on the space from the medulla oblongata to prosencephalon are known to play essential role in the maintenance of the state of wakefulness. Neuronal network ensuring the state of wakefulness is activated by endogenous and exogenous stimuli. Wakefulness is accompanied by fast, low-voltage electrical activity in the cortex and the subcortical structures of the brain and by a significant amount of tonus in the muscular system. When awake, adult humans show low- voltage (10-30μV), fast (16-25 Hz) EEG activity known as *alpha-rhythm.*

Sleep onset is the coordinated process, and it includes simultaneous changes in sensory, motor, autonomic, hormonal and cerebral functions. At the same time, any changes that may be observed with a beginning of sleep, are not exceptionally specific for the state of sleep (McGinty D, Szymusiac R, 1994). Sleep is defined behaviorally by four criteria (Sheldon HS et al., 1992):

(1) reduced motor activity,
(2) decreased response to stimulation,
(3) stereotypic postures (for example, lying down with eyes closed in humans), and
(4) relatively easy reversibility.

Passage from wakefulness to sleep is characterized by progressively slower frequencies and higher voltage activities in the EEG. At the beginning of sleep, EMG can register gradual reduction in a muscular tone, however it is not always observed. Perceptive abilities of the person can markedly vary. Thoughts become fragmented, although most studied persons believe that they do not sleep, even when the first stage of sleep can be objectively documented in them (Kryger MH et al., 1989). Ability to perform elementary tasks disappears with transition to sleep, and an ability of a person to maintain some physical activity several seconds after the appearance of the low-voltage activity of the mixed frequency on EEG is considered as an example of so called automatic behavior. Impaired perception of the surrounding, which is observable at the beginning of sleep, may be demonstrated by the lack of response to visual and acoustic stimuli and occurrence of this reaction only when EEG comes back to a pattern specific for the sate of wakefulness. At the same time, an ability of the CNS to integrate stimuli remains after the beginning of sleep, since the threshold of awakening is markedly lower for the significant irritants than for the insignificant ones. It is common that at the transition from the state of wakefulness to the state of sleep, a person can experience fragmented visions. Generalized or focal involuntary muscular contractions (jerks) called *hypnotic myoclony* may be observed in wake-sleep transition (Kryger MH et al., 1989). These myoclonic events may become more evident, if the person has been stressed or has experienced changes in daily routine. Transition from wakefulness to sleep can be accompanied by amnesia (Drucker-Colin RR, McCaugh JL, 1977).

There are two distinctive sleep phases. One is known as *orthodoxical, or quiet, or non-rapid-eye-movement (non-REM, NREM) sleep*, the second is so-called *paradoxical, or non-quiet, or rapid-eye-movement (REM) sleep* (Sheldon HS et al., 1992). Healthy children of the older age groups and healthy adult humans usually fall asleep by entering non-REM sleep, which is identified primarily by its EEG characteristics. Sleep starts with slow, rolling eye movements and low voltage, mixed-frequency EEG activity. In this stage there is some activity of skeletal muscle but no rapid eye movements. This is designated as stage 1 of non-REM sleep, which lasts only several minutes and is the transition from wakefulness to the onset of sleep. Stage 2 of non- REM sleep is known as the bursts of 12-14 Hz sinusoidal waves called sleep spindles and high-voltage biphasic waves called K complexes occurring episodically against a background of continuing low-voltage EEG activity (Figure 1). When high-amplitude, slow (0.5-2 Hz) delta waves appear in the EEG, stage 3 can be scored. At the stage 4, slow-wave activity increases and dominates the EEG record. Stages 3 and 4 in the humans are sometimes called slow- wave sleep (SWS). Overall, neuronal activity is at a low level during these phases. This is reflected in metabolic rate and brain temperature, which are at their lowest. Thus, sympathetic outflow is decreased as heart rate and blood pressure decline. The pupils are constricted as parasympathetic activity increases and then dominates.

REM sleep is an active form of sleep. The REM sleep phase is characterized not only by rapid eye movements, but also by essentially complete inhibition of skeletal muscle tone (atonia), and, in some animals, EEG patterns similar to those obtained during wakefulness. Most dreams are thought to occur during this phase of sleep. In the REM phase, the EEG reverts to a low-voltage pattern similar to stage 1 non-REM sleep. Because the EEG during REM sleep closely resembles that of wakefulness in some animals, this phase has also been called paradoxical sleep (Figure 2). Indeed, during REM sleep the discharge patterns of most neurons resemble those during active wakefulness, and certain neurons — those in the pons, in the lateral geniculate nucleus, and in the occipital cortex — even fire in more intense bursts than normally seen during wakefulness. These bursts generate high-voltage spike potentials, the ponto-geniculo-occipital or *PGO spikes* named for the structures in which the spikes appear most prominently (the pons, lateral geniculate nucleus, and occipital cortex).

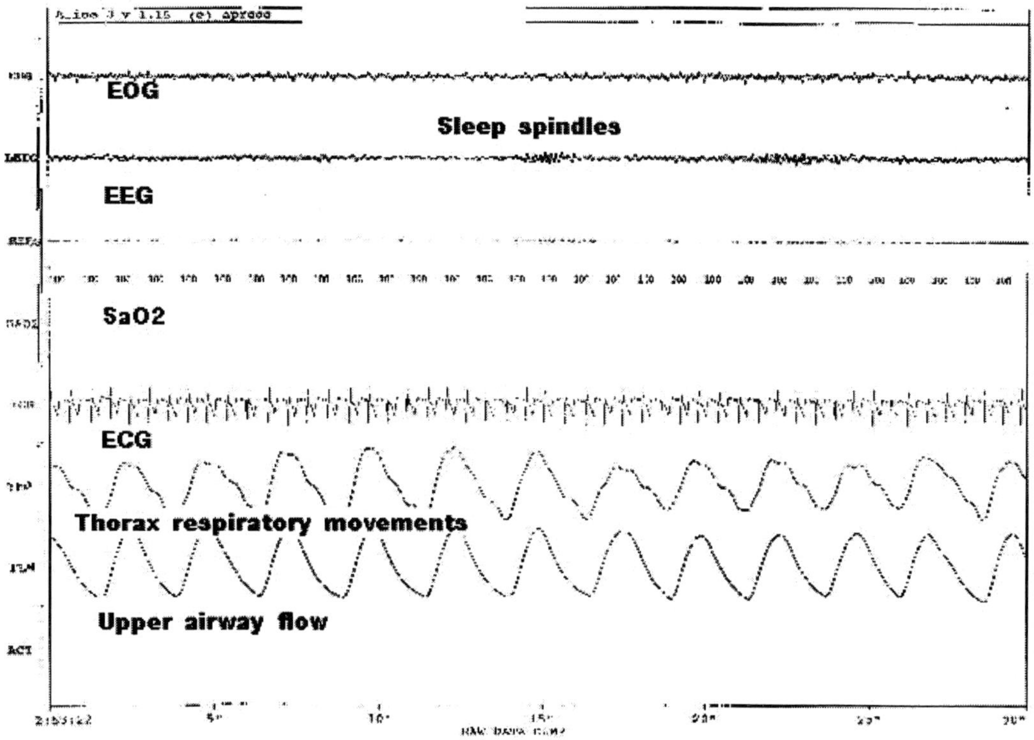

Figure 1. Polysomnographic picture of sleep stage 2 in infant

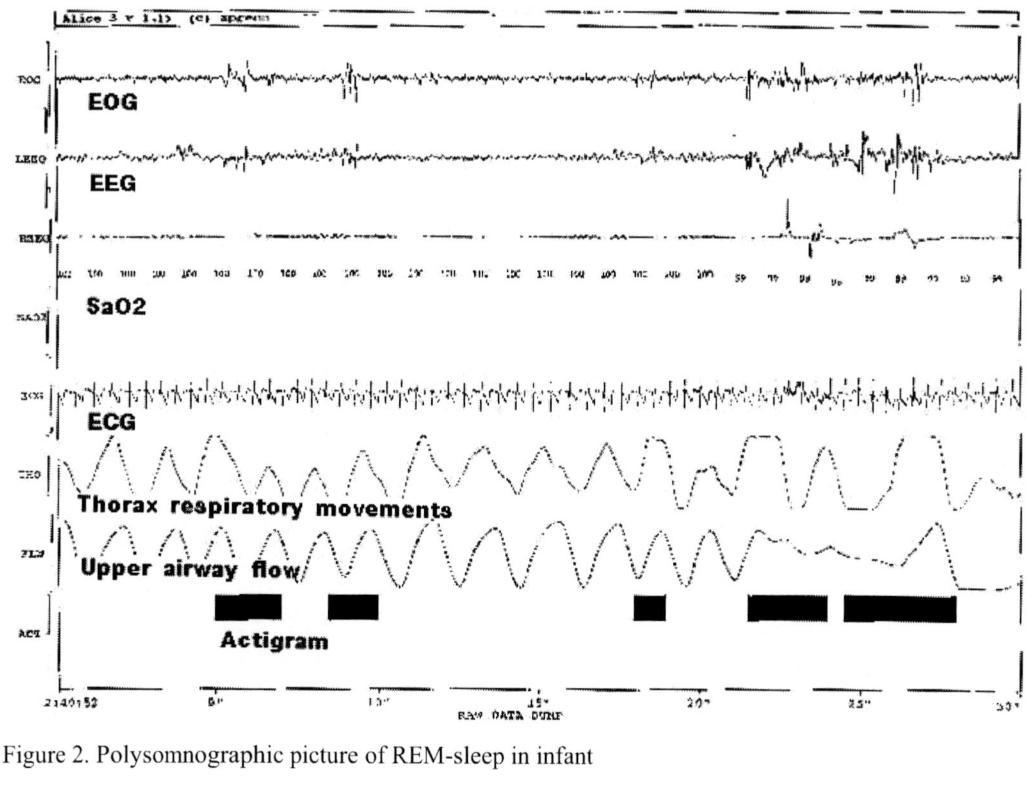

Figure 2. Polysomnographic picture of REM-sleep in infant

PGO-like waves can be evoked in alert waking by abrupt stimuli, similar to those that elicit startle, suggesting that the spontaneous PGO spikes of REM sleep may be generated by internal activation of the neural circuit for the startle response. Saw-tooth waves are also typical EEG pattern of REM sleep in the frequency range 2 to 6 cycles per second with characteristic appearance (slow negative upward phase, followed by a steep positive deflection). The source and function of saw-tooth waves are unknown, theyoccur intermittently, frequently in association with REM bursts, and can also occur in stage 2 sleep just before REM sleep.

The phenomena in paradoxical sleep may be classified under two heads*:* *tonic* (those having to do with continuous phenomena) and *phasic* (those of a periodic character*). The principal tonic phenomena* observed are PGO spikes, very regular "theta" waves at the level of the hippocampus (a structure running from the front to the rear of the brain), total disappearance of electrical activity in the muscles of the neck, bursts of eye movements and cardio-respiratory irregularity. *The principal phasic phenomena* are high voltage spikes, isolated or grouped in volleys that appear at the level of the pons and the rear part of the cortex (which is associated with the visual system). These spikes make their appearance about a minute before the tonic phenomena. Just as the latter show up, the peripheral phasic phenomena come into evidence.

Consistent with the overall increase in neural activity during REM sleep, brain temperature and metabolic rate are high, equal to or greater than during the waking state. Surprisingly, however, almost all muscle tone is lost (atonia); small, phasic twitches do occur and the skeletal muscles controlling the movements of the eyes, middle ear ossicles, and diaphragm do not become atonic. Myoclonic head jerks may also occur. In REM sleep, the pupils are at their most constricted (miosis), reflecting the high ratio of parasympathetic to sympathetic output to the pupil. Penile erections regularly occur during REM sleep. Finally, this sleep stage is also characterized by a reduction in homeostatic mechanisms. Respiration is relatively unresponsive to changes in blood CO_2, and response to heat and cold are absent or greatly reduced. As a result, body temperature drifts toward ambient temperatures.

These observations make it clear that sleep cannot be easily placed in a continuum from light to deep. Each phase is behaviorally complex, with its own configuration of differentially-activated physiological mechanisms. By some criteria REM sleep might be considered lighter than non-REM. e.g., humans are easier to awaken from REM than from non-REM stages 3 and 4. By other criteria non-REM might be considered lighter than REM sleep, e.g., muscle tone, spinal reflexes, and body temperature regulation are maintained during non-REM, but are reduced during REM.

Non-REM and REM phases alternate cyclically during sleep (Sheldon HS et al., 1992). Human adults usually begin sleep by progressing from stage 1 through stage 4 of non-REM sleep. This progression is intermittently interrupted by body movements and partial arousals. After about 70 to 80 minutes the sleeper usually returns briefly to stage 3 or stage 2 and then enters the first REM phase of the night, which lasts about 5 to 10 minutes. The length of the cycle from the start of non-REM sleep to the end of the first REM phase is about 90-110 minutes. Typically, this non-REM-REM cycle is repeated 4 to 6 times a night. In successive cycles the duration of stages 3 and 4 decreases while the length of REM phases increases. In young adults REM phases constitute 20-25% of total sleep time; stages 3 and 4 non-REM about 15-20%; and stage 1 non-REM about 5%. The largest amount of sleep time, 50 to 60%, is spent in stage 2 non-REM sleep (Figure 3).

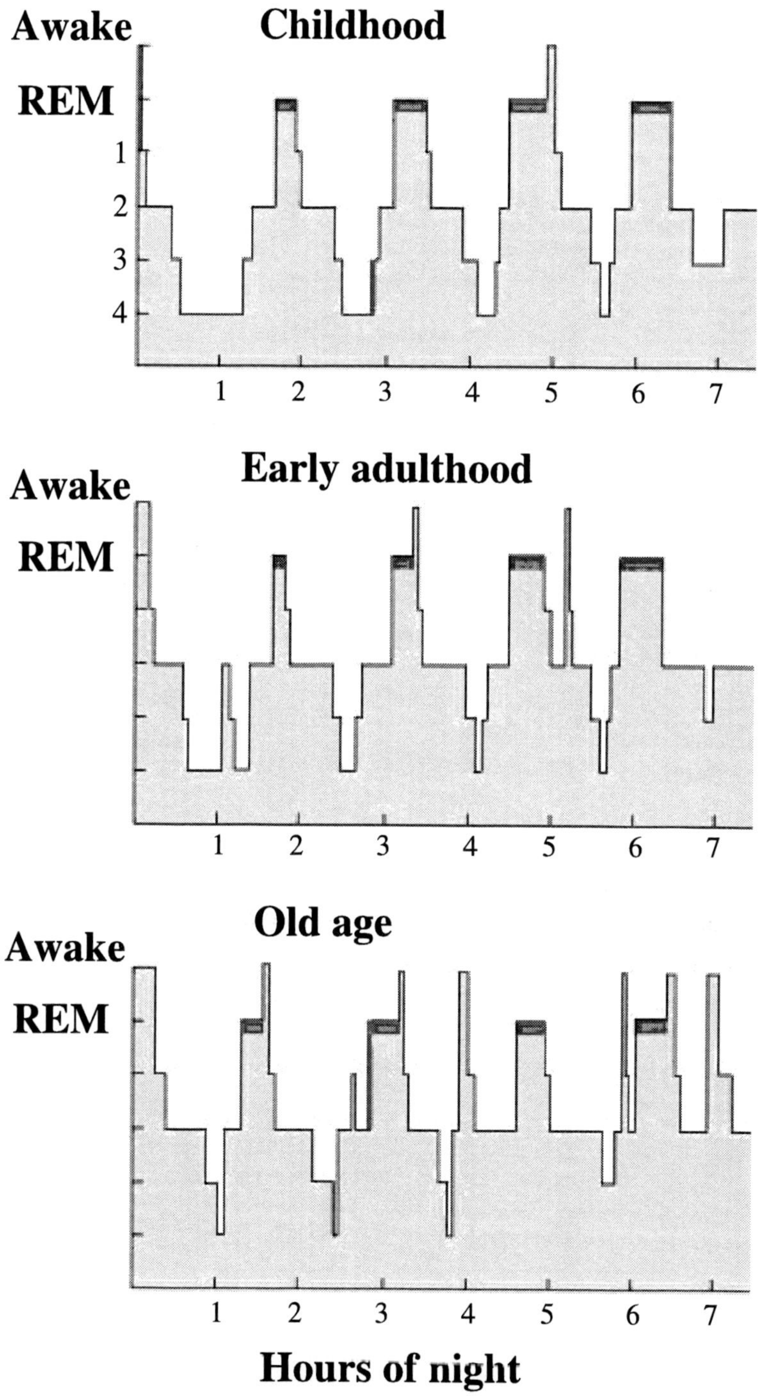

Figure 3. Schematic representation of the cycling of human sleep stages at different times of life

The sates of wakefulness and sleep are caused by the activity of exciting and inhibiting neurons located in various sites of the brain stem and forebrain. These neurons are organized as the systems, or the "networks", each involved into control of the certain state. Two major groups (or neuron systems) participating in the origins and alternations of the episodes of SWS and REM-sleep have been described (Hobson JA, 1974):

(1) Aminergic system ("REM-off" cells) localized in the dorsal raphé nuclei (serotoninergic), locus coeruleus (noradrenergic) and n. peribrahialis lateralis (noradrenergic). The cells of this system are characterized by maximal bioelectric activity during wakefulness. Their ability to generate electric potentials progressively decreases during SWS, and becomes minimal in REM-sleep.
(2) Cholinergic reticular system ("REM-on" cells) which is mainly localized in mesencephalic, medullar and giant-cell tegmental pontine area. In the experimental animals, it has been found that the activity of the above neuronal structures is reciprocated to the activity of aminergic " REM-off " cells: the minimal activity is marked in a sate of wakefulness, it is increasing at transition to SWS and is reaching its maximum in the REM-sleep. The cells of the given system are cholinoceptive and cholinergic.

Two systems cooperate, and this leads to alternations of SWS and REM-sleep. Therefore, monoaminergic system plays a role in regulation of the sleep stages, first of all the phase sleep phenomena observed in the REM-sleep (Jouvet M, 1994; Lydic R, Baghdoyan HA, 1994). As well, these systems participate in numerous complex regulatory mechanisms related to modulation of the functions connected with sleep, such as thermoregulation, an ability to learn, appetite, etc. Many neurons and glial cells can produce, transport and allocate biochemical substances that may change neuronal bioelectric activity and induce sleep onset.

The neuronal machinery most important for REM sleep resides in the nucleus reticularis pontis oralis which lies in the caudal midbrain and the rostral pons. Bilateral destruction of this nucleus eliminates REM sleep for extended periods. Microinjection of the acetylcholine agonist carbachol in the same portion of the pons elicits extended periods of REM sleep in experimental animals. Thus many of the neurons critical for REM sleep are responsive to acetylcholine. The most common cell type in the brain stem has maximal levels of activity in both REM sleep and active waking and minimal rates during non-REM sleep (*REM-waking-on cells*). Some cells with this discharge pattern have projections to motoneurons in the spinal cord and extraocular muscle nuclei. Burst firing in these neurons mediates movement of the head, neck, limbs and eyes during waking, as well as the rapid eye movements and muscle twitches that occur during REM sleep despite peripheral motor inhibition. The PGO spikes characteristic of REM sleep originate in a group of cholinergic neurons in the pons. These *PGO cells,* fire in a burst that initiates the PGO spike in cells in the lateral geniculate nucleus of the thalamus. It is thought that these bursts are due to a low-threshold calcium spike in the pontine PGO related cells, similar to the low threshold calcium spikes involved in the generation of sleep spindles. Destruction of these cells blocks PGO activity, while other aspects of REM sleep continue. Conversely, stimulation of this area produces PGO waves, even in the absence of REM sleep. *Serotonergic neurons* of the raphé nuclei, a group of cells that lie along the midline of the brainstem, have a *REM-off* discharge pattern. These cells have an important role in regulating the discharge of the cholinergic cells responsible for PGO spikes. During waking serotonin blocks the burst firing mode of PGO cells by hyperpolarizing them. In the transition from non-REM to REM sleep the cessation of activity in serotonergic cells allows the PGO cells to begin discharging in bursts, generating PGO waves. *Noradrenergic neurons* in the locus coeruleus and *histaminergic neurons* in the posterior hypothalamus have a *REM-off* discharge pattern similar to that of the serotonergic cells in the raphé nuclei. Cessation of activity in serotonergic, noradrenergic, and histaminergic cells during REM sleep may contribute to changes in autonomic tone, EEG, and other state-specific changes in neuronal activity. Cessation of activity in the noradrenergic cells of the locus coeruleus and serotonergic cells of the raphé during REM sleep is due to the release of GABA onto these cells by populations of *GABAergic REM-on neurons*. The location and driving force behind the activity of these GABAergic cells remains to be determined. It is likely that further analysis of the conditions altering the excitability of REM-on and REM-off cells will provide important clues to the functional role of REM sleep (Jouvet M, 1994).

Muscle tone is at its lowest level in REM sleep. This reduction in muscle tone is actively induced by the release of glycine onto the motoneurons. The circuitry mediating this suppression of muscle tone resides in the brain stem. Abnormal activity in the brain stem neurons responsible for suppressing muscle tone during REM sleep is believed to contribute to cataplexy, one of the primary symptoms of narcolepsy, REM sleep behavior disorder, and sleep apnea, which will be discussed in the following book sections. A system triggering muscle atonia is activated in REM sleep leading to the discharge of GABAergic neurons, the inhibition of noradrenergic and serotonergic neurons and the activation of cholinergic neurons in the pons. These cholinergic neurons excite pontine neurons that use glutamate as their transmitter. The glutamatergic neurons project to the medulla, where they terminate on interneurons that release glycine onto motor neurons. This release of glycine hyperpolarizes motor neurons leading to the motor paralysis of REM sleep (Juvet M, 1994).

A pontine system with ascending connections contributes to the production of an activated EEG in REM sleep. Some cholinergic cells and adjacent non-cholinergic cells activated during REM sleep project to GABAergic cells in the thalamus. Release of acetylcholine and other transmitters blocks the burst firing mode of these GABAergic neurons and contributes to the change from a non-REM sleep EEG to one of REM sleep.

The waking EEG is also maintained by depolarization of the thalamic GABAergic neurons. During waking, in addition to the rostrally projecting cholinergic neurons active in REM sleep, rostrally projecting serotonergic, noradrenergic and histaminergic neurons are active and contribute to the creation of the waking EEG.

The third class of cells in the nucleus reticularis pontis oralis are referred to as REM-on cells. These cells show little or no activity in waking and non-REM sleep but have high levels of activity in REM sleep. Although there are relatively few of these cells, they play the key role in REM sleep control. One subtype of these cells is GABAergic and is responsible for the inhibition of activity in serotonergic and noradrenergic cells during REM sleep. Another subtype of REM sleep-on cell is responsible for the loss of muscle tone in REM sleep (Jouvet M, 1994).

Sleep researchers have searched for substances that accumulate with waking and are metabolized in sleep. However, no substance is yet widely accepted as being the driving force for sleep. Among those substances that have been identified as having hypnogenic properties are muramyl peptides (a chemical related to substances found in bacterial cell walls), interleukin-1 (a cytokine that may mediate the effects of muramyl peptides as well as immune responses), delta sleep-inducing peptide (a substance isolated from the blood of sleeping rabbits), and a long-chain fatty acid primary amide, cis-9,10-octadecenoamide. Other naturally occurring substances may be sleep factors. Growth hormone and prostaglandin PGD2 increase both REM and non-REM sleep; arginine vasotocin increases non-REM sleep and suppresses REM sleep; insulin increases non-REM sleep. Several substances produce relatively selective increases of REM sleep, including vasoactive intestinal peptide (this may interact with acetylcholine), cholecystokinin and bombesin (both are released in the gut after food ingestion), somatostatin, corticotropin-like intermediate lobe peptide, and prolactin (Jouvet M, 1994).

Another powerful hypnotic substance facilitating sleep is *melatonin.* Melatonin (N-acetyl-5-methoxytryptamine) is a chronobiotic, a neurohormone involved in the regulation of the sleep-wake cycle. It is produced in the pineal gland through the conversion of tryptophan, first to serotonin, then N-acetylserotonin, and finally to melatonin. Melatonin is normally secreted during darkness, in response to the release of norepinephrine from retinal photoreceptors and the resulting activation of the retino-hypothalamic-pineal system. Levels begin to rise at nightfall, with peak serum concentrations typically occurring between 2 and 4 AM in adults. Secretion during daylight is minimal. Melatonin promotes sleep through regulation of the activity of the suprachiasmatic nucleus, located within the anterior hypothalamus above the optic chiasm. Impairment of melatonin production, secretion, or binding within the suprachiasmatic nucleus may lead to chronic disturbances in both sleep onset and duration (Repprt SM et al., 1988; Gordon N, 2000).

Generally speaking, the advent of each sleep stage assumes interaction of two types of neuronal networks. The first is the *executive* network responsible for the sleep phenomenology (any phenomenon characteristic for the certain sleep stage is caused by the activity of the certain neuronal network). The second neuronal network is known as *permissive* which is responsible for the very fact of sleep onset. REM-sleep is formed first in the ontogeny. Reticular formation of pons and a spinal cord play a key role in regulation of the REM-sleep. The executive system the REM-sleep includes "REM-on" neurons specific for each phenomenon characteristic for the REM-sleep (Sakai K, 1988; Jouvet M, 1994). Formation of the SWS also assumes participation of the executive and permissive structures. The activity of executive nonREM-system is identified only by two attributes: by sleep spindles and by the low-frequency, high-voltage EEG. Sleep spindles are mainly caused by the activity of the neurons of thalamic reticular formation. The slow, synchronized electric activity is registered in the various cortical and subcortical regions, however the origin of the typical slow EEG waves characteristic for SWS mostly depends on the integrity of neocortex and is linked with hyperpolarization of the pyramid cells. The isolated thalamic reticular nucleus keeps an ability for the rhythmic electric oscillations. This leads to a suggestion that it is functioning as a pacemaker and is a subject to the influence of permissive system (Steriade M, Buszacki G, 1990). The inhibitory system of a slow-wave sleep includes the neurons sensitive to acetylcholine (pontine and a midbrain nuclei, and also some structures of the forebrain), to histamine (posterior parts of hypothalamus) and noradrenaline (locus coeruleus). Many functions of a slow-wave sleep are under control of the mechanisms providing thermoregulation. The SWS is characterized by reduction in metabolism and body temperature that is associated with increased synthesis of the brain proteins accumulation of the energy reserves necessary for the maintenance the REM-sleep and the state of wakefulness (Jouvet M, 1994).

Chapter III

Sleep Ontogeny

The organization of sleep undergoes marked transformation during ontogeny. Spontaneous motor activity of a fetus comes to light as early as 10 wk of gestation, and cyclic organization of this activity may be recorded from the 20 wk of gestation. At 28-30 wk of gestation, short-term dormant periods of the fetus may be scored, however their cyclic organization is still poorly expressed. With an advancing gestational age, specific cyclic organization of the fetal behavior is formed, which may be considered as the equivalent of alternating states of sleep and wakefulness. Certain stages of such fetal behavior have been designated. *Stage 1 F* is characterized by regular, slow reductions of heart rate, starts ("startles"), absence of eyeball movements. *Stage 2 F* is characterized by irregular heart rhythm, movements of eyeballs, incidental body movements. *Stage 3 F* is known for regular accelerated heart rate, movements of the eyeballs, and the absence of the body movements. *Stage 4 F* can be scored by increased and irregular heart rate, movements of the eyeballs, constant movements of a body of a fetus (Prechtl HFR, 1988). Such cyclic organization of fetal behavior can be found at 30-32 wk of gestation (Nijhuis J et al., 1983; Visser GHH et al., 1987). Studies based on the recordings of the fetal motor activity by the means of ultrasound can reveal stable periods of fetal activity with characteristics similar to what is known for the REM-sleep and SWS; their duration is usually exceeding 3 minutes when the gestational age is 28-31 wk. Obvious correlation between eyeballs fast movements and respiratory movements becomes evident after 27 wk of a gestation (Okai T et al., 1992). Fetal behavior is characterized by certain patterns of complex motor activity that are typical for each stage, and respiratory movements are mostly recorded in *Stage 2 F*. Importantly, fetal behavior at the specified moment of the time does not depend on maternal behavioral condition at this moment (either she is sleeping or awake) (Hoppenbrouwers T et al., 1978).

Clear EEG features of the fetal brain bioelectric activity may be recorded at the 24 wk of gestation. In the pre-term newborn infant, alternating EEG segments of the increased and reduced bioelectric activity may be recorded, which is known as EEG discontinuity, and EEG in pre-term babies born at gestation age less than 30 wk is practically completely represented by such picture. In practically healthy pre-term infant, EEG fits the rule "30-20": periods of low bioelectric activity between the flashes of increased activity are not exceeding 30 seconds at terms of a gestation less than 30 wk; later on, quite intervals between the bouts of increased activity are less than 20 seconds. In the infants who have been born earlier than 30 wk of gestation, EEG can be best recorded in the parietal, central and occipital areas. A concept of *asynchrony* has been put forward. According to this concept, the waves with similar EEG features recorded in the homologous areas (for example, left and right temporal areas) are separated from each other with the time lags equal to 1,5 seconds and more. Pre-term newborns are characterized by various degree of such physiological asynchrony. At the terms of gestation less than 30 wk, the infants are characterized by "hyper synchronization" as a result of the marked immaturity of the brain cortex, and physiological asynchrony appears after 30 wk of gestation and is noticeable up to 36 wk (Lombroso CT, 1985). In the pre-term infants, EEG may reveal the waves of low and middle amplitude with the frequency equal to 10-20 Hz occurring against a background of low-frequency (0,3-1,5 Hz) high-amplitude (50-250 μV) delta-waves. Such EEG pattern is known as *"brushes"*. At the gestational age less than 28 wk, these brushes may be recorded in the central areas, less often in occipital and temporal areas. The brushes may be asymmetric or asynchronous, but quite often symmetric character. In the infants born at term, brushes are seldom revealed in quiet sleep. Another EEG feature noticeable in pre-term babies is monorhythmical alpha-or theta-activity that may be recorded in the occipital area: such feature becomes evident at gestational age less than 28 wk. Such activity lasts 6-10 seconds, and it can be asynchronous or asymmetric. Along with brushes, given EEG feature is the distinctive characteristic of deeply immature babies. With further infant growth and development, episodes of theta-activity with the frequency 4,5-6 Hz can be revealed on EEG in the temporal areas. This form of bioelectric activity comes to light only after 28 wk of a gestation, and it is mostly expressed at 28-32 wk of gestation. In the infants with gestational age less than 28 wk, delta-waves on EEG are usually recorded in the median and central brain areas. Clear delta-activity in another areas of the brain may be determined only after 28 wk of gestation. After 28 wk of a gestation it may be possible to identify the EEG characteristics specific for the SWS. However, SWS is almost negligible in the common sleep structure up to 36 wk of a gestation. EEG features of the SWS are known as so-called *tracé alternant*. This specific variant of bioelectric activity appears at approximately 37-38 wk of gestation, and it becomes the prominent EEG feature of the SWS in the newborn infants. After birth within the first month of life, tracé alternant is gradually replaced by formed sleep spindles. Long sleep spindles (1,8-3,4 sec) may be found in the infants aged 3-4 months. Later on, duration of the sleep spindles is slightly decreasing, reaching the average value of 0,5-0,7 sec by the end of the first year of life. Intervals between sleep spindles are increasing with infant's growth: when the infant is 6 mo old, the length of these intervals is usually equal to 9-11 sec, while at 2 years inter-spindle intervals may be as long as 19-28 sec. Intervals between the spindles are diminishing further with advancing age. True slow waves on the EEG are usually appearing after infant's birth at approximately 8-12 wk (Guilleminault C, 1987), and since then, it is possible to subdivide quite sleep into 4 stages.

It has been stressed that a special terminology for the differentiation of the states in the newborns and young infants should be used, since EEG characteristics and related behavioral features in these babies markedly differ from those observed in the adults and older children. It was suggested to distinguish following major sleep states in early childhood: *active sleep (AS),* or stage 2, and *quiet sleep (QS),* or stage 1. Active sleep is determined by concurrence of fast movement of the eyeballs, increased heart rate variability and increased respiratory rate, reduced muscular tone, numerous body movements accruing against a background of low-voltage EEG. On the contrary, quiet sleep is scored by the absence of fast movements of eyeballs, reduced number of the body movements, increased muscular tone, reduced variability of the respiratory rate and stable high-voltage EEG. Polysomnographic study is an imperative tool for the proper differentiation of the certain sleep sates in the young children. During active sleep, sucking movements, incessant body movements, smiles, grimaces, slight tremor, vocalizations are usually observed. Simultaneously with the phase movements of the eyeballs, bursts of muscular movements and irregular respiration are usually found.

Apart from the specified sleep sates, *indeterminate sleep (IS)* has also been described. This term is applied to the sleep fragments that cannot be ascribed to any of the mentioned above (Anders T et al., 1971). From 27 to 34 wk of gestation, IS makes up about 30 % of the total sleep time. The percent of time falling on IS markedly decreases from the 35 wk of gestation, and remains practically constant until infant's birth (Curzi-Dascalova L et al., 1988).

Apart from the accepted classification of sleep states in the newborns, it has been suggested to classify the states of infant's behavior within as a part of sleep-wake cycle, without regard to the EEG characteristics (Prechtl HFR, O'Brien MJ, 1982):

State 1 - infant's eyes are closed, respiration is regular, there are no body movements;
State 2 - eyes are closed, irregular respiration, no body movements;
State 3 - eyes are opened, there are no body movements;
State 4 - eyes are opened, there are body movements, infant does not cry;
State 5 - eyes are opened or closed, the infant is crying.

The periods of an active and quiet sleep lasting longer than 3 minutes, distinguishable according to EEG patterns and fitting the accepted criteria of certain sleep sates, are evident in pre-term newborns with gestational age exceeding 27 wk without neurological abnormalities (Curzi-Dascalova L et al., 1993). In pre-term newborns, EEG features markedly differ in quiet and active sleep: during active sleep, EEG looks homogeneous, whereas during quiet sleep it is variable. With infant's growth and development, bouts of slow-wave EEG activity are getting longer, and the amplitude of a background EEG observable at the intervals between these bouts is growing higher. As a result, a specific picture of tracé alternant typical for a phase of quiet sleep in mature newborn infant is formed (Eiselt M et al., 1997). Subsequently, during first several weeks of infant's life, tracé alternant is replaced by slow-wave high-voltage EEG, characteristic for quiet sleep.

Duration of sleep states may markedly vary from one sleep cycle to another in one and the same baby. Unlike the adults, pre-term and mature newborns fall asleep in a phase of an active sleep. The first period of an active sleep observable immediately after a condition of wakefulness is usually short, and the EEG observable at this moment has lower frequency compared with the subsequent episodes of active sleep. Duration of sleep cycles is short until 35 wk of a gestation; its mean duration at this stage of infant's ontogeny is about 45-50 min. From the 35 wk of gestation, sleep cycles are lasting 55-65 min, and these values may be found in the infants during first couple of months after birth (Curzi-Dascalova L et al., 1988; Borghese IF et al., 1995).

It is important to note that admission to NICU does not markedly influence sleep structure in pre-term babies without major neurological and organ system abnormalities (Curzi-Dascalova L et al., 1993; Eiselt M et al., 1997). The organization of sleep in the infants who were born pre-term and further reached full post-conceptional age after birth, in the infants born at term with the signs of intra-uterine growth retardation, as well as in the healthy infants born at term is mostly the same (Curzi-Dascalova L, Peirano P, 1989; Scher MS, 1996).

During following age periods, the changes in sleep structure are related to essential modifications of EEG patterns, qualitative characteristics of certain sleep phases, changes of the percent of time falling to certain sleep states and sate of wakefulness. Thus, 3 to 6 wk after birth, tracé alternant disappears on infant's EEG. In the infant born full-term who is 6-9 weeks old, sleep spindles may become identifiable on EEG, and subsequent complete formation of this specific bioelectric phenomenon by the age three months reflects the process of maturing of thalamic structures (Steriade M et al., 1985), and active growth and a myelination of neuronal dendritic spines (Hughes JR, 1996). Identifiable K-complexes also appear on the EEG by the end of the third month of life (Samson-Dolfus D et al., 1977), and it has been suggested that this is connected with further activation of infant's information processing, since maximal ability of an infant to master information is observed after the appearance of K-complexes on EEG (Halasz P et al., 1976). The distinction of four states of quiet-sleep is possible at the age of 1,5-3 months (Sterman MB et al., 1977).

During the first year of a life, gradual increase of a proportion of quiet sleep is observed (Tables 1, 2). Sleep structure undergoes prominent changes during the first 6 mo of life. In particular, the proportion of active sleep decreases from 60 % in the newborn to 34 % at 3 months and 31 % at 6 months. Simultaneously, proportion of quiet sleep is growing up: from 49 % at the age of 3 months to 55 % at 6 months (Curzi-Dascalova L et al., 1988; Hoppenbrouwers T et al., 1988; Curzi-Dascalova L, Peirano P, 1989). After 6 months, further reduction in proportion of active/REM-sleep is observed, and this is mainly connected with diminishing number of the episodes of active sleep, while mean duration of each episode remains stable. By the age 2 years, stable proportion of quiet sleep at night is usually found (Louis J et al., 1997)..

Total sleep duration in infants and children decreases from approximately 16-17 hours in neonatal period to 14-15 hours by 6 mo of life, 13 hours by 2 years, 9 hours by 10 years and 7-8 hours in teenagers (Clement CD et al., 1972). It is worth mentioning that a remarkable inter-individual variability of sleep duration that may exceed 2 hours does exist, however normally in a given infant or child of a given age sleep duration is stable enough (Klackenberg G, 1982). There are interethnic differences of sleep duration in children of the same age group (Ottaviano S et al., 1996). Reduction of sleep duration, which is observed in children during the first 6 mo of life, is mostly connected with a reduction of the length of a day time sleep (Louis J et al., 1997).

After 6 years, reduction of sleep duration is related to the later hours of putting a child to sleep, while morning wakening hours mostly remain the same.

Table 1. Percent of active sleep time (Kelmanson IA, 1995)

Age, weeks	Centiles for distribution					
	Girls			Boys		
	10	50	90	10	50	90
≤ 6	57,9	72,7	89,0	57,3	73,9	88,1
7-8	49,3	65,6	81,3	47,4	63,5	76,5
9-10	45,8	56,9	70,1	47,3	56,9	69,4
11-12	44,8	56,2	68,0	42,3	55,8	68,2
13-14	43,3	55,2	69,1	44,2	54,3	63,8
15-16	44,9	57,2	67,3	44,1	53,9	68,0
17-18	46,1	56,4	70,6	41,0	53,2	63,8
19-20	41,8	52,9	65,8	37,9	50,8	68,4
21-22	43,9	52,9	68,1	39,4	50,9	65,8
23-24	40,6	50,7	62,9	40,1	50,9	66,4
≥ 25	38,1	50,6	62,8	40,4	51,2	65,5

Table 2.Percent of quiet sleep time (Kelmanson IA, 1995)

Age, weeks	Centiles for distribution					
	Girls			Boys		
	10	50	90	10	50	90
≤ 6	9,1	18,6	30,5	7,3	14,1	28,0
7-8	14,9	25,2	37,6	12,8	23,7	32,8
9-10	19,9	28,8	36,4	19,8	28,1	36,5
11-12	21,9	30,8	38,0	22,7	31,2	39,2
13-14	21,8	32,5	41,7	25,5	33,6	41,7
15-16	24,2	34,7	42,3	24,0	34,4	45,1
17-18	23,9	36,3	47,0	27,8	37,7	45,8
19-20	27,1	37,4	45,5	24,6	38,8	47,8
21-22	24,3	40,6	48,5	28,7	38,3	48,6
23-24	32,0	40,4	50,0	28,8	41,2	51,7
≥ 25	30,4	44,2	54,2	28,7	43,3	54,1

An ability of an infant to concentrate is achieved in parallel with development of quiet/NREM sleep and steady sleep patterns. These findings are the arguments that development of inhibitory mechanisms connected with increasing complexity of neural networks and neuro-chemical maturation is a protracted process (Clement CD et al., 1972; Sheldon HS et al., 1992). By the age 3 mo, regular alternations of active and quiet sleep are formed (Guilleminault C, 1987). Until the infant is 3 month old, high concordance between physiological parameters of an organism during sleep may be noticed. The organization of sleep structure is poorly controlled by feedback mechanism. In particular, there is no sufficient modulation of cardiac rhythm and respiration in various sleep states. The latency of active/NREM sleep in infants younger than 3 mo is only 8 minutes or less. By the age 6 mo, entering to sleep through active/REM sleep is only observed in 18 % observations, and most infants usually enter the sleep through SWS, as it happens in the adults. Short-term awakenings are common in infants during first 2 months of life (Guilleminault C, 1987). It is more common for the infants aged 1-2 months to wake up in the morning through the sate of active, rather than quiet sleep.

Until the child is 5 years old, the periods of REM-sleep across the sleep tine have almost same length, irrespective of the child's sleep at the day time. Subsequently uniformity of REM-sleep episodes disappears. The first episode of REM-sleep at night becomes relatively short, whereas the following episodes are getting longer with approaching early morning hours, the same as is observed in the adults. By the age of 5 years, the ratio between quite and REM-sleep becomes the same as is found in the adults, and the volume of body movements in sleep decreases.

Length of certain sleep states and their relationship in ontogeny have important physiological content. Quite sleep, especially SWS, which becomes more prominent with infant's growth and development, is related to energy accumulation, which will be required in the REM-sleep for adequate function of the synapses, cellular mitosis, and liberation of growth hormone (Oswald I, 1980; Adam K, Oswald I, 1983). It was shown that the peaks of pulse-like growth hormone secretion are observed at the beginning of sleep, and up to 70 % of episodes of growth hormone releases coincide with SWS (Holl RW et al., 1991; Van Cauter E et al., 1992). REM-sleep was shown to play important role in maturation and differentiation the CNS, unfolding of genetic programming (Jouvet M, 1986), consolidation of memory and learning (Cipolli C, 1995). Therefore, it is not surprising that antenatal and early neonatal periods are known for high proportion of active sleep in total sleep time. REM-sleep is characterized by high activation the CNS. In the adults, cerebral blood flow, a reliable indicator of the neuronal activity, increases in REM-sleep up to the level characteristic for the state of wakefulness, and often even may exceed that level (Madsen PL, Vostrup S, 1991). Similar findings may be observed in the infants born with gestational age above 36 wk: in these babies, cerebral blood flow is higher in active sleep than in quiet sleep or in a state of wakefulness (Mukhtar AL et al., 1982). At the same time, in the pre-term infants, cerebral blood flow in active sleep is lower than in a quiet sleep (Greisen G et al., 1985).

Association between the frequency and character of rapid eyes movements seen in the REM-sleep on one hand and cognitive activity in this state of sleep on the other has been confirmed for the adults. In this regard, it is interesting to note that the frequency of rapid eyes movements increases during the first 6 months of infant's life, and the trajectory of these movements gradually becomes more complicated during the first year of life (Ktonas PY et al., 1990).

Several hypotheses were put forward explaining possible physiological value of the high proportion of active/REM-sleep in total sleep time characteristic for the early age (Roffwarg HP et al., 1966). First, immature nervous system of the infant and young child is unable to inhibit the influence of brain structures responsible for REM-sleep (Valatx JL, 1998). Thus, reducing proportion of REM-sleep with infant's growth and development may reflect CNS maturation. Secondly, developing nervous system requires stimulation for the natural growth and maturation, and REM-sleep may play a role of that endogenous stimulator. It is logical to assume, that reduction of expressiveness of REM-sleep during an ontogeny reflects reduction of organism requirements in active sleep during the process of CNS maturation. It was shown that the length of quiet wakefulness - condition in which the infants can adequately adjust to surrounding – negatively correlates with the length of REM-sleep during the first months of life (Denenberg VH, Thoman EB, 1981). Attempts to deprive the newborn infants of REM-sleep were unsuccessful (Anders TF, Roffwarg HP, 1973): these attempts resulted in total sleep deprivation and were associated with increased risk of functional impairments, including thermoregulation and feeding problems.

Active sleep in the fetus and in the newborn may serve as a triggering factor for the unfolding of the genetically programmed behavioral features (Jouvet M, 1978). In the newborn infants, facial mimic similar to the smile in older babies is mostly recorded in the state of active sleep. Frequency of the rapid eyes movements finely correlates with the frequency and duration of smiles of infants. It is exceptionally in the sate of active sleep that another infant's facial mimic resembling the mimic in the adults can be recorded (Challamel MJ, 1992).

Chapter IV

Sleep Chronobiology

The objective researches on live organisms revealed cyclic variation of their physiological parameters, or *biological rhythms*. In particular, depending on the period of fluctuations of a studied parameter, it is possible to speak about *circadian* (the period about 24 hours), *circannual* (the period about 1 year), *ultradian* (the period is less than 20 hours) and *circamensual* (the period about 30 days) rhythms (Halberg F et al., 1972; Aschoff J, 1978; Reinberg A et al., 1987). Circadian rhythms are the best studied, and they, as a rule, have relatively high amplitude. For each biological periodic phenomena including circadian rhythms it is possible to reveal "peaks" *(acrophase)* and "troughs" *(batiphase)* at certain time. Distribution of "peaks" and "troughs" for a set of various physiological parameters reflects the time organization of physiological parameters of the organism in a 24-hour scale.

Biological rhythms have genetic origin, they are endogenous, i.e., can persist in the absence of environmental time cues, and their periodicity can be changed only within very narrow limits. Each organism is characterized by presence of the set of oscillators (or internal clocks) which are connected among themselves hierarchically. Nevertheless, the phases of the rhythms are normally maintained in adaptive relationship to the environment by their response to external cues called Zeitgebers ("time givers") (Aschoff J, 1978). Sunlight is a powerful Zeitgeber which can be linked to active or inactive phases of the circadian rhythms. Thus, most adult humans sleep at night, whereas nocturnal animals such as rats and mice sleep mostly when it is light.

One major internal clock or pacemaker of circadian rhythms is the suprachiasmatic nucleus of the anterior hypothalamus. Lesions of the suprachiasmatic nucleus dampen the circadian rhythm of sleep as they dampen other circadian rhythms, and these rhythms can be restored by transplanting a fetal suprachiasmatic nucleus. Entrainment of rhythms to light is mediated partly through a retinohypothalamic tract to the suprachiasmatic nucleus.

Transitions from one condition to another are difficult to reveal in children with gestational age less than 36 wk. Rapid eye movements represent one of the major distinctive characteristics of a rudimentary active sleep at the pre-term infant. This phenomenon is accompanied by a characteristic picture on the EEG only since 30-31 wk of gestation (Curzi-Dascalova L et al., 1988). Since then, occurrence of specific activity, characteristic for an active sleep, is not spontaneous any longer, and has typical cyclic organization, despite immaturity of the brain structures.

Right after birth, circadian rhythms that have originated antenatally, may temporarily disappear, despite sufficient functional maturity of those central structures which regulate circadian rhythms. The organization of sleep-wake pattern in the newborn infant is poorly represented by circadian rhythms and markedly expressed by mixed ultradian rhythms (Glotzbach SF et al., 1995). In the newborn infant, the periods of sleep and wakefulness alternate with ultradian periodicity of about 3-4 hours, and these rhythms have endogenous character and do not depend on a regime and a character of infant's feeding (Salzarulo P, 1980; Scher MS et al., 1992). Such periodicity is known as *basic rest/activity cycle (BRAC)*. Episodes of the REM/active and SWS/quiet sleep cyclically alternate with ultradian periodicity. These features may be observed within the first several weeks after infant's birth. It has been suggested that this may be related to competing interaction between circadian rhythms of sleep-wake cycle and high needs of the newborn infant in active, "paradoxical" sleep; with further reducing expression of the REM-sleep in the total sleep structure, circadian rhythms are restored again.

Generalized motor activity in the newborn infants correspond to the certain stages of the sleep-wake cycle, and it differs in the infants born full-term and pre-term. With advancing postconceptional age, the form of infant's motor activity is becoming more complicated (Robertson SS, 1987). Thus, myoclonic and generalized body movements are mostly observed in the pre-term infant, while localized and segmentary movements dominate in the full-term baby (Prechtl HFR et al., 1979). "Internal clocks" are becoming more organized with increasing influence of the environmental stimuli, in particular alternation of light and darkness, fluctuations of temperature, noises, and social factors.

Specific changes of gross and fine motor activity characteristic for the certain sleep stages occur in parallel with maturation of brain bioelectric activity. In a typical case, the ultradian sleep cycle starts as an active sleep noticeable immediately after the baby is falling asleep; this patter may be seen in 50 % of the newborns. This initial fragment of active sleep is represented by EEG of the mixed frequency occupying from 25 to 30 % of the total sleep cycle duration. Then the short period of high-voltage, low-frequency EEG can be noticed, corresponding to a quiet sleep; the length of this fragment of a quiet sleep is about 3-5 % of the total sleep cycle length. This is followed by EEG picture of what is known as tracé alternant, and the specified fragment makes up about 25 % of a sleep cycle of a newborn infant. The cycle ends up with occurrence of another burst of an active sleep - a fragment of low-voltage irregular EEG waves, occupying about 15 % of a sleep cycle length. IS occupies approximately 10-15 % of a sleep. Episodes of wakefulness are evenly distributed across the daytime and night.

The circadian pacemaker presumably comes into effect during the last months of intra-uterine development. Thus, in the infants born pre-term at 28-34 wks of gestation, circadian rhythms of body temperature and alternations of the episodes of rest and motor activity have been revealed; the periodicity is approximately 25 hours (Mirmiran M et al., 1990). Circadian rhythms are finally developed within the first several weeks of infant's life (Tomioka K, Tomioka F, 1991). By the end of the first month of life, duration of night-time sleep becomes longer than in the day time, and it is in the early morning hours that the prolonged period of infant's wakefulness is observed.

It has been suggested that activation of the circadian rhythms in a fetus is markedly dependent on the degree of fetal maturity, and in the infants born pre-term, the organization of microenvironment that will promote formation of the circadian rhythms may accelerate the rate of their development and maturation (Mann NP et al., 1986; Fajardo B et al., 1990).

During first three months of life, infants born full-term will experience reorganization of the alternations of the periods of sleep and wakefulness, and circadian periodicity becomes more expressed in them. By the end of the second month, infants usually sleep at night, and by the 3-rd month of life 70 % infants have consolidated night sleep lasting about 5 hours. By that time, increased production of melatonin during night sleep may be clearly detected. The most essential changes in sleep duration are observed in the infants during the first three months of life. It in at this period of postnatal life that active myelination of the neuronal dendritic spines of the forebrain, active synaptogenesis, increasing interactions between brain stem and thalamo-cortical structures normally take place (Van der Knaap MS, Valk J, 1990). At this stage of infant's growth, ultradian periodicity of the episodes of sleep and wakefulness become similar to the adult features. By the age 3-4 months, synchronization of the periods of sleep and wakefulness with 24-hour cycle becomes obvious (Coons S, 1987). By the age of 6 months, consolidated periods of night sleep lasting approximately 7 hours are normally formed (Anders TF et al., 1992). The longest periods of sleep and wakefulness may be observed in the infants at the fixed points around the clock, and the longest periods of sleep are observed immediately after the longest periods of wakefulness (Coons S, 1987).

During first several months of life, circadian rhythms of the heart rate, body movements, body temperature, secretion of cortisol and a melatonin are formed, and the amplitude of fluctuations of these parameters markedly increases after the age of 3 months (Lodemore M et al., 1991; Glotzbach SF et al., 1994; Weinert D et al., 1994; Sadeh A, 1997). Apparently, these rhythms play essential role in formation of the cyclic organization of the episodes of sleep and wakefulness and influence emergence of the certain sleep stages in the infants (Sadeh A, 1997).

Circadian periodicity of sleep/wake cycles and their synchronization with micro-environmental factors is mainly accomplished during the first year of life, and this process is intimately connected with a degree physiological maturity of infant's organism. During the first weeks of infant's life, tactile sphere (the skin and mucosa) plays the leading role, and infant's regular care for itself serves as the potent synchronizer (feeding, dressing, bathing, etc.). Beginning with the second month of infant's life, after visual and acoustical spheres become more mature, alternations of light and darkness, level of noises in the infant's environment and personal contacts (faces, sounds, speech) start to play more important role.

Consolidation of sleep at night time (settling) is defined as a continuous sleep of an infant from the midnight up to 5 o'clock in the morning (Anders TF et al., 1992). This phenomenon is mostly formed in the infants aged 3 to 6 months, however many infants in whom consolidation of a night sleep has been duly formed may later present with increased frequency of night awakenings after 9 months (Anders TF et al., 1992). Perhaps, such feature may be caused not only by the influence of the social factors (particularly, by micro-environmental), but in many respects may depend on biological factors, and the age of 9 months is known for increased frequency of infant's night awakenings (Louis J et al., 1997).

From the biological point of view, sleep consolidation is associated with maturation of the system producing melatonin. In a 18-weeks-fetus, the suprachiasmatic nucleus of hypothalamus is already sensitive to production of melatonin (Reppert SM et al., 1988). Further maturation of the mechanisms responsible for melatonin production occurs in the first 6 months of postnatal life (Kennaway DJ et al., 1992), and it is at this age interval that the process of night-time sleep consolidation in infants is maximally expressed. Close relationship between the character of melatonin production in the 6-8-month-old infants and the features of the sleep-wake cycle is observed, and the insufficient peak level of melatonin is associated with fragmentation of the night sleep. Similar results were found in the older children (Cavallo A, 1993). Disorders of sleep consolidation are connected with night awakenings in the infants, which is the most frequent form of sleep disturbances in this age group. Normally, most babies get the ability to "sleep at night " by the end of the first year of life. Objective studies with actigraphic recordings during sleep have shown that, on average, normal one-year-old-infants wake up 2 times per night, but they are able to calm down and quickly fall back asleep, often without any parental intervention (Sadeh A et al., 1991). However, the infants with sleep disorders may have on average 4 awakenings per night, and they usually require parental interventions before they fall asleep again. It was reported that 20-30 % infants and young children have the problems connected with frequent night awakenings (Johnson M, 1991). Process of sleep consolidation is continuing after the first year of a life, and therefore some cases of night sleep disorders may further disappear. By the end of the first year of life, the longest episode of sleep can reach 8-9 hours (Anders TF, Keener M, 1985). The longest episode of night sleep reaches 10 hours in 5-years-old children and remains stable until adolescence, then starts to decrease. Despite general tendency to the natural consolidation of the night sleep with infant's growth and development, there are evidence that sleep disorders in the infants may progress, if not properly and duly corrected (Salzarulo P, Chevalier A, 1983). Night awakenings in the infants and young children is a common phenomenon. Such situation may be observed in 20-40 % children aged 1-2 years (Adair R et al., 1991). These physiological awakenings are mostly observed in the REM-sleep (Schulz H et al., 1985). After 6 months, the number of these awakenings essentially decreases, and they mostly become noticeable in the time interval from midnight to 5 o'clock in the morning. In 3- year-old children, night awakenings may practically disappear, possibly due increased length of the sleep cycle and growing proportion of the slow-wave sleep in first half of night.

Unlike dynamic changes during the first year of life, changes of the circadian rhythm of sleep-wake cycle at the age 2-5 years are less sharp. There is further consolidation of sleep in the form of the long night sleeping period and several (usually 2) shorter day time periods: first is usually observed in the morning, the second - at the early afternoon hours. Further the morning day-time sleep disappears, and by 3-5 years it is only night sleep that may be retained (Weissbluth M, 1995a).

Inconsistent data have been provided regarding ontogeny of the sleep latency. It has been reported that the infants and young children have longer sleep latency (Beltramini A, Hertzig

M, 1983), and until preschool age, sleep latency gradually decreases (Ottaviano S et al., 1996).

The studies connected with an artificial manipulation with sleep duration (reduction or increase in its length, shifts of the beginning of sleep), have shown that the magnitude with which quiet sleep, particularly slow-wave sleep, is expressed, positively correlates with the length of the preceding period of wakefulness (Achermann P, Borbély AA, 1990). This has been related to the hypothetical homeostatic *process S.* In the adults, this process S is manifested by the predominating slow-wave sleep at the beginning of sleep and its diminishing intensity with each subsequent sleep cycle (Borbély AA, 1982). Ontogeny of the process S has not been extensively explored so far. Meanwhile, it is known that slow-wave sleep becomes dominating in the first half of the night when the baby is about 3-month-old. It is at this age that slow-wave sleep is becoming maximally expressed during the first sleep cycle. After 6 months the longest period of sleep occurs after the longest period of wakefulness (Salzarulo P, Fagioli I, 1992; Schechtman VL et al., 1994).

Chapter V

The Mechanisms Responsible for the Maintenance of Respiration

The maintenance of normal respiration assumes participation of following basic elements: *the receptors* perceiving the information and transmitting it in the central regulator; *the central respiratory controllers* located in a brain, and *the effectors* (respiratory muscles and the lungs) providing ventilation (West JB, 1985).

The afferent input into the central system is provided primarily by *four groups of neural receptors:*

(1) peripheral arterial chemoreceptors;
(2) central (brainstem) chemoreceptors;
(3) intrapulmonary receptors; and
(4) chest wall and muscle mechanoreceptors.

The *peripheral arterial chemoreceptors* consist of the carotid and aortic bodies. The physiologic significance of the aortic bodies in humans is difficult to determine but likely to be small: the carotid bodies appear to have preeminent importance. The carotid bodies are located at the junction of the internal and external carotid arteries, receiving their blood supply from branches of the external carotid artery, and their sensory supply from the carotid sinus branch of the glossopharyngeal nerve. The carotid bodies consist of two different cell types, glomus cells (type I) and sheath cells (type II). Both cells are innervated primarily by the carotid sinus nerve, which contains both parasympathetic and sympathetic neurons with both afferent (sensory) and efferent (motor) components. The afferent neurons terminate on glomus cells. There is also an unmyelinated supply to the sheath cells. The efferent neurons or sympathetic innervation primarily supply the vasculature and regulate vessel tone and blood supply to the carotid body. The carotid body has a uniquely high arterial blood supply (2 L/min/100 g) that allows for its oxygen needs to be met by the dissolved oxygen in the blood, unlike other tissues that depend primarily on oxygen initially bound to hemoglobin. In addition, PaO_2 is the specific signal sensed by the glomus cells. Therefore, the carotid body is insensitive to conditions that lower arterial oxygen content but exert little or no change on the PaO_2. The carotid body responds to both PaO_2 and hydrogen ion concentration ($[H^+]$). The intensity of the response of the glomus cells varies according to the severity of the arterial hypoxemia or acidosis in a nonlinear manner. The greatest increase is seen in response to hypoxemia, especially when PaO_2 falls in the adults and older children to 70 mm Hg, at which point the firing frequency and, subsequently, minute ventilation increase in an accelerated fashion. This increase in minute ventilation is manifested primarily by an increase in the depth of breathing (tidal volume) rather than by an increased respiratory rate. In mammals, the carotid bodies are responsible for about 90% of the ventilatory response to hypoxemia; the remaining 10% is from the aortic bodies. They are also responsible for 20 to 50% of the response to arterial hypercapnia and acidemia, with the remaining 50 to 80% of the response to hypercapnia and acidemia mediated by the central brainstem receptors (Gonzales C et al., 1992). The exact mechanism by which the carotid bodies transduce the stimulus of a low PaO_2 into neurotransmitter release is still under investigation. An ability of the carotid and aortic chemo-receptors to respond to hypoxia is markedly reduced in the newborns compared with the adults and older children; it varies in postnatal period, mainly during the first two weeks of life. These changes in receptors sensitivity occur in parallel with age-dependent changes of blood oxygen saturation (Blanco CE et al., 1984; Kumar P, Hanson MA, 1989; Williams BA et al., 1991). Nevertheless, hypercapnia is a potent stimulator of these receptors already in the late fetal period, and the reaction of receptors to hypercapnia is well expressed in the newborns.

There are *chemoreceptors in the brain* that are sensitive to CO_2 or hydrogen ions (Bruce EN, Cherniak NS, 1987). Although no definite chemoreceptors have been defined anatomically, results of experiments involving the local application of chemical, electrical, and thermal stimuli or the ablation of neural tissue suggest that the central chemoreceptors are located at or near the ventral surface of the medulla. Stimulation of these receptors increases both the rate of rise and the intensity of the inspiratory "ramp" signal, thereby increasing the frequency of the respiratory rhythm. The inspiratory ramp signal is a nervous signal transmitted to the respiratory muscles as a weak burst of action potentials that steadily increases in a ramp-like manner (Guyton AC, Hall JE, 1996). Physiologically, these central chemoreceptors respond primarily to alterations of $[H^+]$ in the cerebrospinal fluid (CSF) and medullary interstitial fluid. They are located in many different brainstem locations, including the ventral medullary surface, deeper sites near the nucleus tractus solitarius, and rostrally close to the locus coeruleus. Stimulation of these sites increases the whole respiratory control system output, suggesting that they respond to changes in brainstem pH via a ventilatory feedback loop. Hydrogen ions enter and exit the CSF and extracellular fluid in the vicinity of the central chemoreceptors as a result of both CO_2 dissociation and by direct diffusion into and out of the bloodstream.

The ventilatory response to an increased $PaCO_2$ is divided into an initial rapid phase (within seconds) due to the relatively rapid acidification of the CSF, and a slower phase (within minutes) due to hydrogen ion buildup in the more highly buffered medullary interstitium. In addition, when compared with the highly membrane-permeable CO_2, hydrogen ions in the arterial blood penetrate the blood-brain barrier relatively slowly (minutes to hours). As a result, changes in pH result in relatively slower and smaller changes in ventilation. Changes in $PaCO_2$ alone have a potent acute effect on ventilation but only a weak chronic effect. This occurs because a chronically elevated $PaCO_2$ is associated with renal compensation with the retention of HCO_3^-. This HCO_3^- gradually diffuses through the blood-brain barrier and into the CSF, where it binds to the excess hydrogen ions produced by the elevated $PaCO_2$ and negates their effect on ventilatory drive (Bledsoe SW, Hornbein TF, 1981).

The central hydrogen ion chemoreceptors are somewhat redundant, sensing the same stimulus as the carotid bodies. Carotid body chemoreceptors seem to adjust for breath-to-breath changes in arterial chemistry, while the hydrogen ion-sensitive medullary chemoreceptors sustain ventilation with lesser breath-to-breath variation (Bledsoe SW, Hornbein TF, 1981).

An important role has been attributed to opiate receptors of ponto-medullar zone and endogenous endorphin-like peptides (Santiago TV, Edelman NH, 1985). In particular, much interest has been raised in so-called substance P, effective neurotransmitter and a neuro-modulator. Apart from the other physiological effects, substance P provides respiratory regulation (Scholle S et al., 1990). This substance increases the intensity of the functioning of the respiratory center and raises its sensitivity to the afferent vagal influences. The central answer to acute or chronic hypoxia in the newborn infants is extremely unstable due to immature central control: initially, increased ventilation may be marked, but this does not last for a long time (Noble R, 1992).

Pulmonary receptors are present in the airways and lung parenchyma. They are all innervated by the vagus nerve, with myelinated fibers supplying the airway receptors and unmyelinated C fibers supplying the lung parenchyma. The airway receptors are subdivided into the slowly adapting receptors (SARs), also known to be the pulmonary stretch receptors, and the rapidly adapting receptors (RARs). The lung parenchyma receptors are called juxtacapillary receptors. The SARs lie among the airway smooth muscle and are responsible for the Hering-Breuer inflation reflex. Activation of these receptors also causes the tracheobronchial smooth muscle to relax, thus dilating the airways. The Hering-Breuer reflex is the prolongation of expiratory time and the decrease in respiratory rate in response to lung inflation. It is manifest only at a VT of > 3 L and seems to play a protective role in preventing excessive lung inflation (Trippenbach T, 1994). The SARs do not accommodate to persistent stimulation such as prolonged distention and participate in ventilatory control by prolonging inspiration in conditions that reduce lung inflation, such as airway obstruction or decreased chest wall compliance, thereby allowing a normal tidal volume to be achieved. Conversely, in conditions that prolong expiration, lung deflation is slow and the increased SAR activity increases the force of contraction of the expiratory muscles and also prolongs expiratory time. These receptors are particularly important for respiratory control in the newborn infants. The RARs lie between the airway epithelial cells and are irritant receptors, responding to noxious stimuli such as dust, cigarette smoke, and histamine. They are concentrated in the carina and primary bronchi, and are also believed to be cough receptors. These RARs are innervated by myelinated fibers and have a more rapid rate of adaptation than SARs. During normal breathing, their discharge is independent of the phases of inspiration and expiration, and therefore these receptors do not seem to be an important influence on breathing at rest. Their most important function may be to detect pathophysiologic changes in the airways. RARs also seem to be responsible for the augmented breathing and sighs that occur sporadically during normal breathing that helps to prevent atelectases. The pulmonary parenchymal receptors are innervated by the unmyelinated C fibers of the vagus nerve. They are called juxtacapillary receptors because of their location near capillaries in the alveolar walls. Stimulation of these C fibers may also be associated with rapid shallow breathing, bronchoconstriction, and increased airway secretions.

Mechanoreceptors are sensors that respond to changes in length, tension, or movement. The primary mechanoreceptors in the chest are the muscle spindle endings and tendon organs of the respiratory muscles and the joint proprioceptor. Muscle spindles are primarily influenced by changes in length and are responsible for reflex contraction of the skeletal muscles in response to stretching. Afferent information from these receptors is carried in the anterior columns of the spinal reticular pathway and terminates in the region of the respiratory centers in the medulla (Mitchell RA, Berger AJ, 1981). Muscle receptor afferents are involved in the level and timing of respiratory activity. These receptors may also play a role in the increase in ventilation occurring during the early stages of exercise (Duron B, 1981). Tendon organs sense changes in the force of contraction exerted by the muscles of respiration. Tendon organ receptors are involved in monitoring the force of muscle contraction and have an inhibiting effect on inspiration. They may be important in coordinating respiratory muscle contraction during breathing at rest or with a respiratory load. Joint proprioceptors sense the degree of chest wall movement and may also influence the level and timing of respiratory activity. Proprioceptor afferents project to the phrenic motor neurons and affect their firing rate. They also ascend to the medullary respiratory neurons in the dorsal respiratory group and nucleus retroambiguus, where they affect the timing of inspiration and expiration (Duron B, 1981).

The central respiratory controllers are divided into the brainstem group (involuntary) and the cerebral cortex group (voluntary). The former is further subdivided into pneumotaxic, apneustic, and medullary centers.

The *pneumotaxic* center consists of the nucleus parabrachialis and the Kolliker-Fuse nucleus in the pons. This center is important in influencing the timing of the inspiratory cut-off by providing a tonic input to the respiratory pattern generators located in the inspiratory center. Thus, this center may modulate the respiratory response to stimuli such as hypercapnia, hypoxia, and lung inflation, and is of importance in regulating the duration of inspiration (Mitchell RA, Berger AJ, 1981).

The *apneustic* center is found in the lower pons and seems to function as the source of impulses that terminate inspiration, an "inspiratory cut-off switch." Inactivation of this center results in apneustic breathing, which is rhythmic respiration with a marked increase in inspiratory time and a short expiration phase. Apneustic breathing can be experimentally induced in animals by transecting the spinal cord between the pneumotaxic center and the lower brainstem in combination with bilateral vagotomy. Clinical information from patients with brainstem pathology and clinical pathologic correlations suggest that the pneumotaxic center may function similarly in man.

The *medullary* center is divided into an inspiratory dorsal respiratory group (DRG) of neurons and an expiratory ventral respiratory group (VRG). The DRG is located in the nucleus of tractus solitarius in the medulla and is important in integrating impulses from visceral afferents from the upper airways, intra-arterial chemoreceptors, and lung parenchyma through the V, IX and X cranial nerves, respectively (Berger AJ et al., 1977). It may also be the site of projection of proprioceptive afferents from the respiratory muscles and chest wall. The DRG is a processing center for respiratory reflexes and is the site of origin of the normal rhythmic respiratory drive consisting of repetitive bursts of inspiratory action potentials. The exact mechanism by which this rhythm is generated is unknown.

A theory to explain the breathing rhythmicity of the central pattern generator is as follows. Inspiration begins by the abrupt removal of inhibitory impulses to the DRG. This is followed by increased inspiratory motoneuron activity in the form of a slowly augmenting ramp of signals that is suddenly terminated by an off-switch mechanism. This termination of inspiratory activity occurs when a critical threshold is attained. During early expiration, there is another burst of inspiratory neuronal activity that is referred to as postinspiration inspiratory activity, which does not prolong inspiration, but rather slows down the rate of exhalation during the first part of the expiratory phase. The neurons that generate postinspiration inspiratory activity may depend on mechanisms that are different from those responsible for the main inspiratory ramp activity.

The VRG consists of both inspiratory and expiratory neurons and is located within the nucleus ambiguus rostrally and nucleus retroambiguus caudally. The VRG innervates respiratory effector muscles through the phrenic, intercostal, and abdominal respiratory motoneurons (Berger AJ et al., 1977). Its output increases with the need for forceful expiration such as in exercise or in any condition of increased airway resistance to breathing.

The *cerebral cortex* plays a role in ventilatory control and can also influence or bypass the central respiratory control mechanism in order to accomplish behavior-related respiratory activity such as cough, speech, singing, voluntary breath holding. In the adults and children of older age, voluntary respiratory regulation may be achieved due to the cortical control over medullar respiratory centers, and voluntary hyperventilation may lead to such a reduction in arterial pCO2 that respiratory alkalosis may be provoked, sometimes leading to muscular cramps. Voluntary hypoventilation is less easier to be achieved, and the length of voluntary respiratory pauses is determined by numerous factors, including arterial pCO2 and pO2.

Other sites of brain, such as *limbic system* and *hypothalamus*, can influence the mode of respiration, particularly under affective conditions. It has been proved that the medullar respiratory centers not only assure involuntary respiratory drive, but also participate in its voluntary regulation (McKay LC et al., 2003). Attention also has been paid to the role of *cerebellum and vestibular analyzer* in the fine modulation of respiratory rhythm (Harper RM, 2002). Cerebellum contains deep nuclei involved in respiratory control. Activation of the neurons of a fastigial nucleus leads to increased ventilation due to higher respiratory rate and (or) augmented respiratory volume. Destruction of this nucleus does not seriously interfere with the respiratory rhythm under physiological conditions, however may markedly reduce reflex answer to hypercapnia and hypoxia. This is an argument that the neurons of the specified cerebellar nucleus are sensitive to the blood levels of CO_2 and H+. This nucleus is under the inhibiting influence of the Purkinje cells localized in the medial site of the worm of cerebellum, and their stimulation can lead to inhibition of respiration. The neurons of fastigial cerebellar nucleus have chemoreceptive property, and they are not involved in realization of the reflexes connected with mechanoreceptors. Besides, this nucleus is involved in vagus-mediated reflexes related to respiration (Xu F, Frazier DT, 2002). Vestibular analyzer and cerebellum play important role in regulation of a respiration and control of hemodynamics under extreme conditions, and their modulating effect is most salient expressed during sleep. Damages of the vestibular analyzer may lead to disappearance of the typical variability of autonomic characteristics that is normally observed in the REM-sleep (Harper RM et al., 2000), and cerebellar hypoplasia is often associated with major impairments of respiratory control in infants and children (Gonzalez de Dios J et al., 1995).

The *effector system* consists of those pathways and muscles that are involved in the actual performance of inspiration and expiration.

The spinal pathways connect the respiratory centers in the brain and spinal cord to the respiratory muscles and are divided into the descending and ascending pathways. The descending pathways connect the DRG and VRG to the ventrolateral columns of the spinal cord, the phrenic nerves, and the inter-costal and abdominal muscles of respiration. These pathways are used for the inhibition of the expiratory muscles during inspiration and inhibition of inspiratory muscles during expiration to prevent opposing muscles from contracting at the same time (Berger AJ et al., 1977). The ascending pathways connect the respiratory muscles to the higher brainstem levels. Impairment of the ascending spinal pathways may lead to respiratory dysfunction in the form of apnea during sleep that is reversed by arousal. This condition may be caused by damage of ascending spinoreticular fibers, lessening brainstem reticular formation activity and causing a general depression of respiration (Berger AJ et al., 1977).

The respiratory muscles consist of the diaphragm and the intercostal, abdominal, and accessory muscles of respiration. The diaphragm is responsible for the majority (75%) of air movement during quiet inspiration, while the inter-costal, abdominal, and accessory muscles (sternocleidomastoid and other neck muscles) account for the remainder (Ganong WF, 1993). In the infants, the function of diaphragm is almost entirely directed to the maintenance of respiration (Muller N et al., 1979). Inspiration at rest is active and expiration is a passive event in the adult patients with normal lungs. During exercise or in patients with airway obstruction, both inspiration and expiration become active, with expiratory contraction of the abdominal wall and internal inter-costal muscles (West JB, 1990).

The thoracic cage of a newborn is characterized by marked pliability, which is important for the infant during delivery: high thoracic compliance facilitates passage of a fetus and alleviates discharge of the fluid contained in the fetal lungs due to mechanical compression of thorax during delivery. However, after birth such increased thoracic pliability may have negative impact on infant's respiration. In the newborn, thoracic compliance is 3 times higher that of the lungs (Papastamelos C et al., 1995). This leads to the situation that during inspiration an infant has a paradoxical retraction of the thorax. As a results, the efforts required for the maintenance of respiration are increased, and this is most obvious in the REM-sleep, against a background of a markedly reduced muscular tone. Costal and vertebral ossification starts intrauterine and proceeds till 25 years, leading to gradual strengthening of the thorax. Thoracic compliance becomes equal to a pulmonary compliance by 2 years (Papastamelos C et al., 1995), however paradoxical movement of the thorax during inspiration may be observed in healthy infants in the REM-sleep at least until 31 months (Gaultier C et al., 1987). If for some reasons upper airway are obstructed in the infants, paradoxical thoracic movements in sleep may be readily noticed. The shape of the rib cage also undergoes modifications during child's growth and development. In the newborns and in the infants, the ribs are located horizontally leading to the limited reserve for expansion on inspiration and increased risk of respiratory failure in pulmonary and heart diseases. Thus, in the infants under one month old, the share of thoracic cage in the maintenance of respiratory volume during sleep is only 33% (Hershenson MB et al., 1990), whereas with advanced age it reaches about 70% (Tabachnick E et al., 1981). As soon as the child starts to walk, the redistributed muscular tone helps to form an adult-like thoracic shape (Openshaw P et al., 1984).

An important contributor to the normal respiratory function in the infants is the upper airway patency. *Upper airway (UAW)* includes extra-thoracic segment of trachea, larynx, pharynx, nose and mouth. The upper airway is a complex structure that is outlined by the soft tissues forming the nasopharynx and oropharynx within the skeletal boundaries of the mid and lower face. It undergoes morphologic change as the result of growth of the skeleton and surrounding tissues as well as functional changes throughout childhood . All can impact on the size, shape, and stability of the upper airway, particularly during sleep when a significant decline in ventilatory drive and respiratory muscle tone occur. Laryngeal anatomy in the infants and young children of early age is such that the epiglottis contacts the soft palate (Tonkin S, 1975). On one hand, this is beneficial from the standpoint of effective sucking in infants. Meanwhile, this may predispose to UAW obstruction during inspiration since nasopharynx in the infants is in a state of partial obstruction. It has been previously suggested that nasal breathing is obligatory in infancy, but it was further demonstrated that infants are able to breathe through the mouth, although nasal breathing is preferable and more efficient (Miller MJ et al., 1985). To adjust for the reduced functional residual capacity of the lungs and higher propensity to athelectases, a special mechanism called *laryngeal braking* (active narrowing the larynx on expiration) is realized in the infants till the age of 6-12 months.

The lymphoid tissue of the UAW increases from the moment of birth till approximately 12 years. Simultaneously, the carcass of the UAW is becoming more firm. The relationship between adenoid and nasopharynx dimension along the sagittal plane during childhood was found to be relatively constant between 1 yr of age and puberty (Fujioka M et al., 1979). Generally steady growth rate of the total nasopharynx area in girls and boys between 3 and 11 yr of age was also found with mild decrease in nasopharyngeal airway area in both groups between 3 and 5 yr that was parallel with mild accelerated growth of the soft tissues between 3 and 5 yr in males and between 3 and 6 yr in females. For both groups, nasopharyngeal airway area continued to increase linearly thereafter, up until 11 yr of age with no further increase in nasopharyngeal soft tissue area (Jeans WD et al., 1981). The mid and lower face skeleton continues to grow linearly along the sagittal and axial planes throughout childhood and the soft tissues defining the upper airway, including the adenoid and tonsils, grow in constant proportion to the skeletal structures during this period. The maintenance of proportional growth among these tissues ensures airway patency throughout childhood and probably contributes to airway stability in normal children, particularly during sleep (Arens R et al., 2002). In the boys, the larynx increases in sizes and changes the shape at the puberty (Kahane JC, 1982). It has been reported that the size of tongue in the puberty and in post-pubertal age is larger in the boys than in the girls (Kerr WJS et al., 1991). Men have a significantly longer pharyngeal airway (choanae to epiglottis) than women, even when corrected for body height. Furthermore, the modeling analysis demonstrated that increasing airway length is associated with an increase in pharyngeal collapsibility (Malhotra A et al., 2002). Endoscopic techniques confirmed that the retro-palatal airway was more compliant in the men than in the women (greater change in area during inspiration), both during wakefulness and sleep (Rowley JA et al., 2002). It is likely that sexual differences in the laryngeal structure developed in the adolescence can, at least in part, explain increased risk of obstructive apneas in male adults (Guilleminault C et al., 1989), while in the young children these differences in obstructive events are not so markedly expressed (Redline S et al., 1999). Meanwhile, data suggested that obstructive and mixed apneas were more common in the boys than in the girls even in the infancy (Kato I et al., 2000).

Unlike the nasal and tracheal parts of the UAW, which are formed by bones and cartilages (in particular, the wall of trachea contains U-shaped cartilage rings), the pharyngeal part of the UAW does not have internal rigidity, as its walls are made up of the soft tissues. Besides, the pharynx and nose are those two sites of the UAW where the resistance to air flow on inspiration are the highest. During inspiration, intra-pharyngeal pressure depends on the air flow speed and the nasal resistance: the higher are the above, the lower is the intra-pharyngeal pressure. Nasal resistance is markedly variable, depending on the changes of the nasal mucosa congestion and activation of the muscles of the nasal alae (Anch AM et al., 1982; Strohl KP et al., 1982).

The cranio-facial structure has a cephalo-caudal gradient of growth. The mid face has more postnatal growth potential than the cranium, and the mandible has more growth potential than the mid-face. This growth has both vertical and antero-posterior components. In the newborn, the face is dwarfed by and tucked below the neuro-cranium. It must grow down and forward (Enlow DH, 1990). As the maxilla grows, the bony structure elongates and is displaced forward in the antero-posterior plane and downward in the vertical plane. The mandible follows the same pattern. The net effect is one of the continually enlarging airway as the soft palate is carried away from the posterior pharyngeal wall and as the tongue moves forward and inferiorly. The airway should be continually enlarging. The net result, however, is complicated due to the soft tissue growth, as described above.

The breadth of the middle cranial fossa determines the latero-medial dimensions of the pharynx. The mandibular rami, which are well within the zygomas in the early infancy, grow laterally, so that in older children and in the adults mandibular rami are equal to the distance from the lateral edge of one middle cranial fossa to the other.

Development of the brain and the neurocranium control the depth and width of the naso-pharyngeal and oro-pharyngeal airway. The depth of the pharynx and nasomaxillary complex is a projection of the anterior cranial fossae. The breadth of the pharynx is a projection of the breadth of the middle cranial fossae. This naso-maxillary development is linked to development of the neuro-cranium, since the mid-face is attached to the neuro-cranium.

In humans, the spinal cord comes off the brain at an angle that is somewhat greater than 90°. This allows bipedal activity with the face looking forward. A measurement of this flexure is the angle of the antrerior cranial base formed by lines drawn from the nasion point to sella to basion. When this angle becomes less obtuse and approaches 90°, it correlates with a shortened anterior cranial base and often lack of pharyngeal depth with airway compromise (Goldberg JS et al., 1981; Peterson-Falzone SJ et al., 1981).

The bony resorptive and appositional fields mentioned above are very active in the infant's growing mid-face. Bone is resorbed along with all the inner nasal walls except for the very top of the nasal vault. This result in an increase in the nasal airway. At the same time, the naso-maxillary complex is displacing anterior-inferiorly within the limits of the anterior cranial fossae. This also increases the nasal/naso-pharyngeal airway. Maxillary growth "carries" the soft palate also, further enlarging the airway. Naso-maxillary growth transversely under the projection of the middle cranial fossae also increases the airway dimensions (Ferraro NF, 2000). The resorptive/deposition fields even involve the bony portion of the intranasal septum, which may respond to some minor buckling of the nasal cartilaginous septum. The bone responds to the cartilaginous functional forces, resulting in a bony deviation of the septum.

Mandubular growth in the infants has generally the same vectors as the naso-maxillary complex. However, the posterior maxilla can accept bone deposition as it enlarges without any significant structural changes. The mandibular body elongates by encroaching on the territory of the mandibular rami. This is related with an active resorptive field at the anterior ramus and bone deposition at the posterior ramus. The ramus becomes the mandibular body. This lengthening of the body is important to accommodate the tongue. Vertical mandubular growth is also important, since it carries the tongue base inferiorly away from the soft palate, creating more airway room. The mandible also widens laterally as a projection of the breadth of the middle cranial fossae, and this also improves airway dimension (Enlow DH, 1990; Ferraro NF, 2000).

Pharyngeal and a laryngeal muscles are important in maintaining UAW patency. Approximately 30 pairs of muscle are responsible for the patency of the UAW. Table 3 is the summary of the simplified classification of the pharyngeal muscles involved into pharyngeal patency control, which may be roughly divided into the *dilators and constrictors*. Constrictors are located in the upper, middle and lower parts of the pharynx; dilators are more numerous. At a level of the oro-pharynx (between the top of the epiglottis and the edge of the soft palate), the UAW patency during respiration is assured by contraction of m. genioglossus: the contraction of its fan-like projected fibers leads to specific vanral movement of the tongue with increasing pharyngeal lumen (Brouillette RT, Thach BT, 1979; Wilson SL et al., 1980). Electric stimulation of other muscles of the UAW also promotes stabilization the UAW on inspiration (Roberts JL et al., 1984). The m. genioglossus performs maximal functional role, and the reduction in the activity of this muscle is associated with essential increase in UAW resistance to air flow (Remmers JE et al., 1978).

Itra-pharyngeal pressure during inspiration is below the atmospheric. It means that in the pharyngeal lumen, mostly on the level of hypopharynx, the internal forces are formed directed toward its collapse. The intra-phryngeal pressure necessary for the UAW collapse, called the closing pressure, has not been precisely determined in humans. In experimental animals this pressure was found to be negative (Brouillette RT, Thach BT, 1979), however the studies in infants are suggestive that closing pressure most likely has positive values (Wilson SL et al., 1980). This means that the pharynx is able to collapse on inspiration even under physiological conditions. Any thickening of the walls of the pharynx, including those connected with pathological processes, such as tonsillar hypertrophy and fat deposition, may lead to the narrowing of the UAW that may contribute to its collapse.

Table 3. The pharyngeal muscles influencing upper airway patency (Isono S, 2000)

Segment of the UAW	Muscle	Function	Pair of nerves
Pharynx	Constrictors of the pharynx:		
	Superior	Constrictor	9,10
	Middle	Constrictor	9,10
	Inferior	Constrictor	9,10
Nasopharynx	m. palatoglossus	Nasal breathing	9,10
	m. palatopharyngeus	Nasal breathing	10
	m. tensor veli palatini	Oral breathing	5
	m. levator veli palatini	Oral breathing	9,10
Oropharynx	m. genioglossus	Anterior movement of the tongue	12
Hypopharynx	Suprahyoid muscles	Anterio-cranial movement of the hyoid bone	
		m. geniohyoideus	C1, C4
		m. mylohyoideus	C1, C4
	Infrahyoid muscles:	Caudal movement of the hyoid bone	
		m. sternohyoideus	C1, C4
		m. thyrohyoideus	C1, C4
		m. omohyoideus	C1, C4

At the level of naso-pharynx (from the edge of the soft palate to the end of nasal septum) the role of each muscle in the maintenance of the UAW patency depends on preferred breathing route. Contraction of the m. palatoglossus and m. palatopharyngeus opens the retropalatal airway, allowing for nasal breathing. Thus, the specified muscles are considered UAW dilators during nasal breathing (Mortimore IL et al., 1995; Tangel DJ et al., 1995). At the same time, m. tensor veli palatini and m. levator veli palatini, which elevate the soft palate, are considered to be the dilators during oral breathing.

The patency of the pharynx during wakefulness is assured by the cortical control over UAW muscular tone. A number of factors may lead to reflex activation of these muscles, among them the negative intra-pharyngeal pressure (Horner RL et al., 1991; Van der Touw T et al., 1994; Mortimore IL, Douglas NJ, 1997), increased speed of the air flow through the nose (McNicholas WT et al., 1993), nasal obstruction (Kuna ST, Smickley J, 1988).

Endoscopic inspection has allowed to reveal those sites of the UAW which are characterized by the maximal propensity to a collapse (Isono S et al., 1997; Isono S et al., 1998). Table 4 is the summary of the most collapsing segments in children and adults in the norm and against a background of respiratory disorders.

In clinically healthy infants and children, most collapsing segment of the UAW is oro-pharynx. However, in the adults in more than 80% observations the most collapsing site is the naso-pharynx. Therefore, most collapsing sites of the UAW shift from the caudal to cranial segments in the process of infant's growth and development. The hypopharynx (a site between the vocal cords and a top of epiglottis) and the oro-pharynx are almost the only collapsing segments the UAW in infants and young children.

Table 4. Distribution of the pharyngeal segments according to the maximal collapsing propensity (Isono S et al., 1997; Isono S et al., 1998)

Most collapsible site	Infants and young children		Older children and adults	
	Norm	Pathology	Norm	Pathology
Naso-pharynx				
Adenoids	0	57	0	0
Soft palate	54	29	82	88
Oro-pharynx				
Tonsils	15	14	0	0
Tongue	31	0	18	12
Total:	100	100	100	100

Their protection against collapse during respiration is assured due to simultaneous activation of the muscles of the UAW involved in respiration, swallowing and speech, regulation of thoracic and abdominal pressure, and normally UAW are able to resist the fall in transmural pressure, equal to a difference between atmospheric and intra-phryngeal pressure. At the level of hypopharynx, dilatation of the UAW is provided due to interactions between the muscles attached to the hyoid bone from the above (m. geniohyoideus, m. myohyoideus), and from the below (mm. sternohyoideus, thyrohyoideus, omohyoideus). These muscles influence position of the hyoid bone and the sizes of pharyngeal lumen (Van de Graaf WB et al., 1984; Van Lunteren E et al., 1987).

In both adults and infants, nasal air passages are most likely to be initially blocked during sleep. In response to nasal occlusion, switching from nasal to oral breathing is necessary to maintain ventilation. However, in infants and children this switch occurs much more slowly than in the adults, and in the REM-sleep this switch to oral breathing is markedly depressed. Pharyngeal edema or damage of the receptors may result in impairment of the reflex, especially in cases of sleep-disordered breathing.

At the level of larynx, UAW patency is also regulated by two antagonistic groups of muscles: the adductors, which close the vocal cords, and the abductors, which open the vocal cords. The adductors include the transverse arytenoids, the lateral cryco-arytenoids, the inter-arytenoids and the thyro-arytenoids, all of which are innervated by the recurrent laryngeal nerve. Abduction of the vocal cords is performed solely by the posterior crico-arytenoids, which are also innervated by the recurrent laryngeal nerve branches. Activity of the laryngeal muscles varies with rhythmic alternations inspiration and expiration. During quiet breathing, the glottic aperture is greater during inspiration, which is caused by phasic laryngeal abductors activity, usually starting prior to contraction of the inspiratory pump muscles (England SJ et al., 1982). Phasic expiratory activity in the abductors is likely to counterbalance the tonic expiratory activity of the posterior cryco-arytenoids, leading to active return of the vocal cords towards the midline during wakefulness. During sleep, the tonic activity of the posterior cryco-arytenoid muscle as well as the phasic activity of the adductors during expiration disappears, while the posterior cryco-arytenoid muscle preserves its phasic inspiratory activity (Kuna ST et al., 1988). It is likely that inspiratory dilation and expiratory narrowing of the laryngeal aperture is actively controlled by the antagonistic laryngeal muscles during wakefulness, whereas the return of the vocal cords toward the midline during expiration appears to be passive phenomenon during sleep (Kuna ST et al., 1990). In those situations when ventilation is stimulated by hypoxia or hypercapnia, the size of laryngeal aperture increases, and laryngeal resistance to air stream decreases; whereas hypocapnia leads to the narrowing of the laryngeal aperture and increased laryngeal resistance (Bartlett D, 1979). The laryngeal caliber is controlled in relation to the magnitude of respiratory activity. The inspiratory dilation of the larynx seems to be beneficial for efficient breathing, and expiratory narrowing may play a role in braking expiratory airflow while controlling lung volume. Normally, there is no marked limitation of the air stream through the larynx during inspiration, even with a reduced tone of the posterior cryco-arytenoid muscle during sleep. However, airflow limitation is evident when the bilateral recurrent laryngeal nerves are damaged or when laryngospasm is induced. Vocal cord dysfunction in Shy-Drager syndrome is known to produce a distinctive stridor during inspiratory flow limitation. Laryngomalacia in the newborns is another condition that causes inspiratory airflow limitation at increased ventilation.

Numerous reflexes are involved in the regulation of the balance of the forces that determine UAW patency. Stimulation of chemoreceptors due to increased pCO2 level or reduced pO2 leads to activation of the muscles that expand UAW lumen (Brouillette RT, Thach BT, 1980).

Such stimulation seems to be "selective", since activation of the UAW muscles is more expressed than activation of the diaphragmatic muscle. The stimulation of the receptors due to the low pressure in the nasal cavity and larynx during inspiration leads to the rapid stimulation of the dilator muscles of the UAW (Mathew OP et al., 1982; Hwang JC et al., 1984; Van Lunteren E et al., 1984). This effect is selective for the UAW muscles. Moreover, decreased pressure leads to reduction of the diaphragm activity. These reflexes are most clearly expressed in the infants (Thach BT et al., 1989). The receptors responsible for these reflexes are located on the surface of the mucous layer of the UAW. They may be blunted by the influence of the local anesthetics.

The involuntary reflexes arising in response to stimulation of the laryngeal irritant receptors may be divided into *active* and *passive.* The reflexes ensuring swallowing, coughing and expiration lead to dynamic elimination of the foreign bodies and substances from the UAW and may be ascribed to active. On the contrary, it is possible to provoke reflexes leading to breath holding and laryngeal closure, known as passive reflexes. Active reflexes are typical for the wake sate and during SWS,. However, during REM/non-quiet sleep, passive reflexes are more common, sometimes leading to long apneas (Sullivan CE et al., 1978). In the infants born pre-term, long apneas in response to stimulation of laryngeal receptors may occur much more often than in the infants born full-term (Pickens DL et al., 1989).

One more mechanism of participation the CNS in protection the UAW from an occlusion has been shown in experimental researches on animals: if respiratory ways appeared partially or completely closed, there was the immediate reflex answer directed on reduction of nasal resistance due to inflating of wings of a nose, to ventral displacement of tongue due to activation m. genioglossus and on disclosing of a larynx due to activation of a back crycoarythenoid muscle (Brouillette RT, Thach BT, 1980; Mathew OP et al., 1982; Van Lunteren E et al., 1984).

The local factor influencing UAW patency are the *surface tension forces*. For some time it has been recognized that the pressure required to open an occluded airway is greater than that required to collapse it. In the closed pharynx, there is an adhesion between the mucous layers of the collapsed pharyngeal walls, which may be quantitatively measured, and which aggravates the effect of UAW obstruction (Wilson SL et al., 1980; Reed R et al., 1985). The additional efforts of the pharyngeal muscles directed at the separation of these "glued" walls are therefore required. The increase in weight of the soft tissues of the neck located outside the UAW may increase the effect of the surface tension forces (Koenig JS, Thach BT, 1988). Surface tension may be reduced with surfactant and increased with saline. Correspondingly the opening pressure is increased with saline and decreased with surfactant (Kirkness JP et al., 2003). Surfactant reduces the resistance at peak pressure while saline did not significantly influence resistance. Saline did not significantly alter the sleep characteristics or disordered breathing events (Morrell MJ et al., 2002).

The influence of *lung volume* on the upper airway has been recognized for some time with changes in airway size (Kamal I, 2002) and resistance (Series F et al., 1990) reported with changing lung volume. It has been demonstrated that raising lung volume reduced pulmonary resistance independent of pharyngeal muscles tone or chemical stimuli, and it has been concluded that caudal traction on the upper airway structures may have caused this effect (Begle RL et al., 1990). In experimental animals, caudal tracheal displacement decreased the compliance and closing pressure of the airway and increased the maximal airflow during progressive decreases in pressure at the larynx. As the studies were performed on an isolated airway during general anesthesia with neuromuscular blockade, these effects are likely due to the mechanical effects of displacement as opposed to muscle recruitment (Tuck SA, Remmers JE, 2002a; b). Lung volume may be a determinant of the closing pressure of the airway in anesthetized humans, becoming more positive (the airway becoming more collapsible) with increasing depth of anesthesia. The increase in airway collapsibility is likely due to the decreased lung volume associated with deeper anesthesia as the end expiratory esophageal pressure was reduced with lighter anesthesia (Eastwood PR et al., 2002).

Increased laryngeal resistance in the infants is involved in the fine mechanisms directed at the maintenance of the fixed level of the end-expiratory lung volume (Harding R, 1984; England SJ, Stogren HAF, 1986). Numerous physiological functions, such as coughing, crying will demand narrowing or even full closure of the UAW at the laryngeal level for the maintenance of the raised intra-abdominal and intra-thoracic pressure. Active regulation of intra-thoracic pressure and the maintenance of the lung volume are integrated into those mechanisms that co-ordinate general motor activity in the infants. As a result, marked changes in respiratory rhythm associated with infant's motor activity may be observed. Reduced minute ventilation resulted from the decreased respiratory rate during motor activity may lead to mild hypoventilation, but in some cases this may lead to significant impairments. Thus, apnea and bradycardia may occur against a background of vigorous motor activity in the newborn infants, especially born pre-term. These episodes are highly reminiscent of what is known as *"affective-respiratory attacks"* in the older children, and can be characterized as *"expiratory apnea"* (Southall DP, Talbert DG, 1988).

Chapter VI

Respiratory and other Physiological Characteristics during Sleep

The sleep and its certain phases are accompanied by the marked changes of the major physiological characteristics, including respiratory. At sleep onset, the behavioral and cognitive influences on ventilatory control are largely eliminated. The NREM/quiet sleep, especially delta-sleep, is characterized by reduced, stable metabolic level. Brain temperature decreases, and the cerebral blood flow is reduced, although peripheral vessels are dilated. In the state of NREM/quiet sleep, the respiratory rate is reduced, respiratory rhythm is regular, which is not observed in a state of wakefulness and in the REM/active sleep. Such character of respiratory rhythm is probably related to the reduction of the number of functionally active neurons in the respiratory center. An increase in airflow resistance typically occurs at sleep onset because of relative hypotonia of the upper airway dilatory muscles (Wiegand DA et al., 1990). Minute ventilation is reduced, but the tidal volume increases as a results of prolonged inspiration in this state of sleep. Respiratory responses to exogenous (and presumably endogenous) stimuli such as the response to hypoxia and hypercapnia are generally reduced (Douglas N et al., 1982a; Douglas N et al., 1982b). Hypoventilation during slow-wave sleep is a product of a decreased tidal volume and respiratory rate resulting in a 2- to 7-mm Hg increase in $PaCO_2$ and a reciprocal fall in PaO_2. A periodic breathing pattern following sleep onset commonly occurs in stages 1 and 2 of sleep. The respiratory pattern is notably most regular in stages 3 and 4.

Ventilatory compensation to both added and intrinsic resistance to breathing is severely reduced during REM/active sleep. Another notable change related to that sleep state is the marked reduction in skeletal muscle tone, sparing only the diaphragm and ocular muscles. This decreased muscle activity is thought to be caused by REM-related supraspinal inhibition of the alpha motor drive and the specific depression of fusimotor function, causing attenuation of the stretch and polysynaptic reflexes with decreased muscle tone. The diaphragm is relatively spared because it contains very few muscle spindles and is therefore much less affected by these inhibitory impulses. However, during phasic REM sleep, the contraction of the diaphragm becomes uncoordinated (so-called "diaphragmatic fragmentation") (Hendricks JC, Kline LR, 1991). In fact, respiration during phasic REM sleep is very irregular and consists of sudden changes in both respiratory amplitude and frequency, which are linked to the rapid eye movements. SaO_2 is also lowest during this period. Ventilation during REM sleep is primarily maintained through diaphragmatic contraction. It follows that any respiratory impairment in which diaphragmatic dysfunction is can cause severe nocturnal hypoventilation, especially during REM sleep. Despite constriction of the vessels of the skeletal muscles, cerebral blood flow in a tonic phase is essentially increased. Brain temperature is also raised, indirectly reflecting increased metabolism. The thermoregulation during REM/active sleep is practically absent; there are neither perspiration nor contractile thermo-genesis, and the body temperature becomes close to the ambient. Reaction of the central and peripheral chemo-receptors to the reduction in pO2 and increased pCO2 is markedly reduced. The state of REM/active sleep is characterized by changes of neuronal activity of the medullar respiratory centers, directly connected with the rapid eye movements and muscular jerks. It allows to suggest that respiratory activity in the REM/active sleep is to a greater extent supervised by the central mechanisms, rather than related to metabolic needs and function of chemo-receptors. Table 5 is the comparison of the major respiratory characteristics in the REM/active and NREM/quiet sleep.

Associations between respiratory patterns and certain sleep stages are evident already in the newborn infants. At the age of 1 wk, respiratory rate is higher during active than during quiet sleep (Finer NN et al., 1992). This increase in respiratory rate leads to increased minute respiratory volume. During the first months of infant's life, inspirations and expirations take more time during quiet sleep. In healthy infants, the tidal volume and the length of respiratory cycle decrease in parallel with increasing number of the rapid eye movements characteristic for the active sleep. Variability of the respiratory cycle length and of the tidal volume are 2-4 times higher in the active than in the quiet sleep.

Table 5. Respiratory activity in REM/active and NREM/quiet sleep (Sheldon HS et al., 1992)

NREM/quiet sleep	REM/active sleep
1. Respiratory rate is less than in the state of wakefulness or in the REM-sleep	1. Tidal volume is less than in the wake sate or in the NREM sleep
2. The maximal speed of the air flow is less than in a state of wakefulness	2. Supplementary muscles involved in respiration are hypotonic or atonic
3. Increased UAW resistance	3. UAW muscles are hypotonic or atonic
4. Increased tidal volume	4. Arrhythmic breathing, apneas and hypopneas may be observed
5. Minute ventilation is decreased	5. Reduced response to hypoxia and hypercapnia
6. Reduced response to hypercapnia	6. Unstable and variable reaction of the laryngeal muscles and diaphragm to UAW occlusion
7. Reduced response to hypoxia	7. Respiration is not controlled by chemo-receptors, vagal and thoracic afferent impulses

Changes in breathing patterns during a sleep in response to chemical stimulation markedly depend on infant's age and maturity. Unlike marked stimulation of respiration seen in the adults in response to hypoxia, suppression of breathing in response to reduced pO2 level may be observed in the fetus. Full-term newborn infants have two-phase reaction to hypoxia: first, increased ventilation is noticed in them, but it returns to the baseline level or may even fall bellow that. Although chemo-receptors are already active in fetus and in the newborn infant, central neuronal synapses may be still immature, therefore hypoxic suppression of the CNS can overweigh afferent carotid body stimulation (Haddad GG, 1986). Fall in the level of blood oxygen saturation as low as 70-75 % does not awoke arousal reaction in the infants and young children.

The specific phenomenon normally observed during sleep in the infants and young children are the respiratory pauses - *apneas*.

There are three types of apneas that can be distinguished clinically and by PSG study (Kelmanson IA, 1997):

(1) *Central apnea (CA)* when at the end of expiration respiratory movements and the air flow through the UAW disappear. Synchronous absence of thoracic respiratory movements and air flow through the mouth and nose is recorded (Figure 4).
(2) *Obstructive apnea (OA)* when air flow through the UAW is blocked due to the collapse of UAW, although thoracic respiratory movements remain (Figure 5)
(3) *Mixed apnea (MA)* when episodes of UAW obstruction are partly associated with the absence of thoracic respiratory movements and partly with their presence. Polysomnographically, this phenomenon is presented with the episode of central apnea, immediately followed by obstructive apnea (Figure 6).

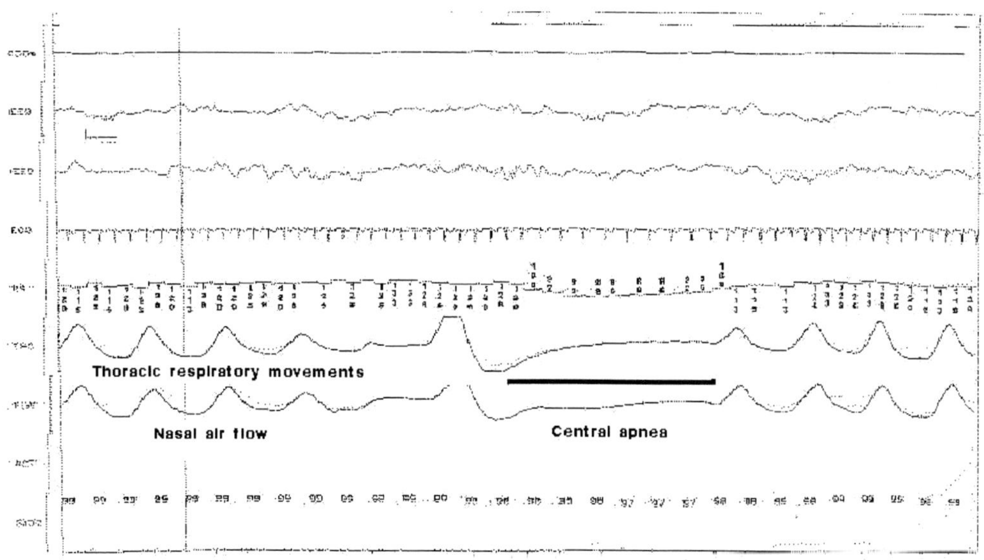

Figure 4. Polysomnographic picture of central sleep apnea in the infant

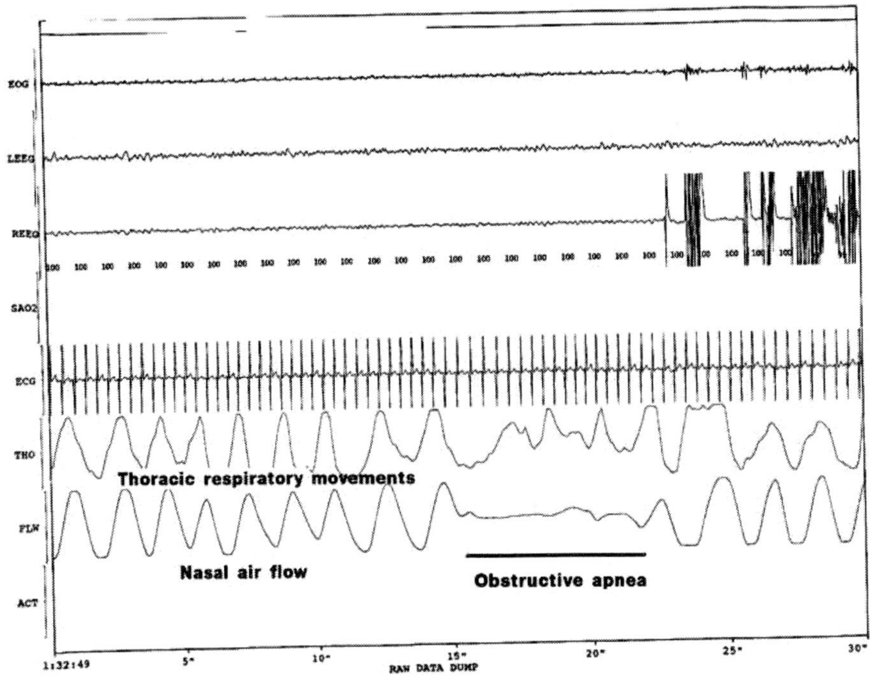

Figure 5. Polysomnographic picture of obstructive sleep apnea in the infant

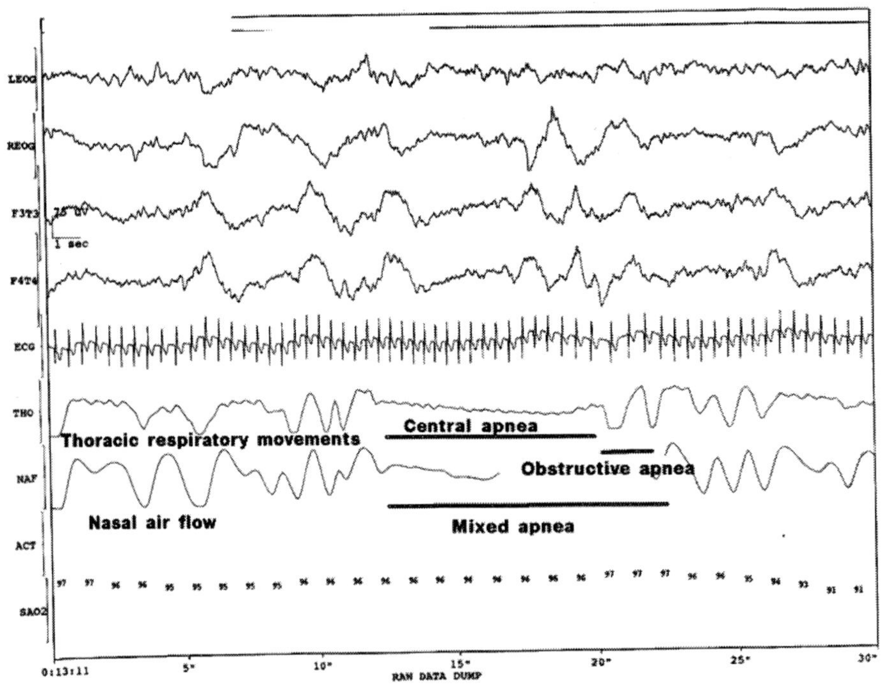

Figure 6. Polysomnographic picture of mixed sleep apnea in the infant

There are common pathways in the origins of CA and OA. In particular, both CA and OA are accompanied by the loss of the UAW muscular tone, primarily m. genioglossus, and UAW collapse is marked in 50 % cases of CA in the infants (Milner AD et al., 1980). This may, at least in part, explain the opportunity of combination of the central and obstructive apneas in the form of mixed apnea.

CA is the common phenomenon in infants and young children (Tables 6-15). Their frequency is increased in the infants born pre-term, both in the REM/active and NREM/quiet sleep.

Table 6. Number of central apneas per 1 hour of total sleep time (Kelmanson IA, 1995)

Age, weeks	Centiles for distribution					
	Girls			Boys		
	10	50	90	10	50	90
≤6	1,1	7,3	20,0	3,2	7,7	28,1
7-8	2,2	6,0	13,4	1,7	4,3	15,9
9-10	2,0	5,4	15,4	1,7	5,0	11,1
11-12	1,8	5,4	12,4	1,7	5,1	10,4
13-14	2,0	5,2	10,3	1,9	4,1	8,4
15-16	1,7	4,8	8,8	1,7	4,6	10,3
17-18	1,6	4,0	9,6	1,0	3,8	8,0
19-20	2,9	4,1	8,4	1,6	4,1	7,6
21-22	1,4	4,5	8,3	0,9	3,9	11,9
23-24	1,9	5,0	8,7	1,7	3,9	9,4
≥25	1,5	4,2	8,0	1,3	3,7	6,5

Table 7. Number of central apneas per 1 hour of quiet sleep (Kelmanson IA, 1995)

Age, weeks	Centiles for distribution					
	Girls			Boys		
	10	50	90	10	50	90
≤6	0	1,7	6,7	1,2	3,0	8,4
7-8	0,4	2,1	7,7	0	2,4	6,9
9-10	0,4	2,7	7,1	0,4	2,2	5,8
11-12	0,5	2,8	6,5	0,4	2,8	6,2
13-14	0,4	2,3	5,6	0,6	2,0	5,4
15-16	0,6	2,3	4,7	0,3	2,0	5,3
17-18	0,4	1,6	4,8	0,1	1,6	4,5
19-20	0,8	2,1	3,5	0	1,8	5,6
21-22	0	1,4	3,8	0	1,6	4,5
23-24	0,6	2,4	4,6	0,3	1,8	4,1
≥25	0,4	1,7	3,8	0,3	1,6	3,9

Table 8. Number of central apneas per 1 hour of active sleep (Kelmanson IA, 1995)

Age, weeks	Centiles for distribution					
	Girls			Boys		
	10	50	90	10	50	90
≤ 6	1,6	8,4	22,5	3,4	9,3	31,0
7-8	2,8	7,2	16,4	2,6	5,8	16,5
9-10	2,5	6,1	18,5	2,3	6,1	13,4
11-12	2,6	6,5	15,9	2,1	6,4	13,0
13-14	2,6	7,0	13,4	2,6	5,3	10,8
15-16	2,6	6,2	11,1	2,0	5,5	13,0
17-18	1,9	5,5	11,6	1,2	5,0	11,3
19-20	3,8	6,0	10,8	2,3	5,5	9,9
21-22	2,2	5,8	12,2	1,0	5,4	13,6
23-24	2,5	7,5	12,5	2,6	5,6	12,5
≥ 25	2,4	6,4	12,0	2,0	5,1	9,4

Table 9.Maximal length of the central apnea during sleep, sec. (Kelmanson IA, 1995)

Age, weeks	Centiles for distribution					
	Girls			Boys		
	10	50	90	10	50	90
≤ 6	7,8	11,0	15,0	8,9	11,8	17,2
7-8	7,4	10,0	12,2	7,5	9,0	15,6
9-10	7,3	9,5	12,5	7,0	9,5	12,5
11-12	7,5	9,5	12,0	7,0	9,5	13,0
13-14	7,5	9,5	12,0	7,0	9,5	12,0
15-16	7,3	9,5	12,0	6,8	9,5	12,0
17-18	7,0	9,5	12,4	7,0	9,0	12,4
19-20	8,0	10,0	14,0	7,0	9,5	13,2
21-22	7,0	9,0	11,7	6,9	9,5	13,7
23-24	6,5	9,0	12,5	7,0	9,5	12,7
≥ 25	7,5	10,0	13,5	7,5	9,5	13,5

Table 10. Maximal length of central apnea during quiet sleep, sec. (Kelmanson IA, 1995)

Age, weeks	Centiles for distribution					
	Girls			Boys		
	10	50	90	10	50	90
≤ 6	5,2	8,0	11,4	7,3	9,5	15,5
7-8	5,7	8,0	10,4	5,5	8,8	11,5
9-10	5,5	8,5	11,5	5,5	8,0	11,5
11-12	6,0	8,5	11,5	6,0	8,5	12,0
13-14	6,0	8,5	11,0	5,0	8,5	11,2
15-16	5,5	8,3	11,4	5,4	8,5	11,5
17-18	5,4	8,0	10,4	5,5	8,3	11,0
19-20	5,5	8,5	12,1	5,5	8,5	12,0
21-22	4,6	7,5	11,0	6,2	9,0	11,5
23-24	5,0	7,5	11,5	6,0	8,5	10,7
≥ 25	5,5	8,0	11,0	5,5	8,5	11,5

Table 11. Maximal length of central apnea during active sleep, sec. (Kelmanson IA, 1995)

Age, weeks	Centiles for distribution					
	Girls			Boys		
	10	50	90	10	50	90
≤ 6	7,8	11,0	14,6	8,0	11,0	16,1
7-8	6,4	9,8	12,2	7,0	9,0	15,6
9-10	6,5	9,0	11,5	6,5	9,0	12,0
11-12	7,0	9,0	11,3	6,5	9,0	12,0
13-14	7,0	9,0	11,5	7,0	8,5	11,0
15-16	6,8	8,5	11,3	6,5	9,0	11,5
17-18	6,6	9,0	11,9	6,0	8,5	11,5
19-20	7,0	9,3	12,1	7,0	9,0	13,0
21-22	7,0	8,5	11,7	6,0	9,0	12,3
23-24	6,5	8,8	12,1	6,5	9,0	12,1
≥ 25	7,0	9,5	13,3	7,0	9,0	12,5

Table 12. Mean length of central apnea during quiet sleep, sec. (Kelmanson IA, 1995)

Age, weeks	Centiles for distribution					
	Girls			Boys		
	10	50	90	10	50	90
≤ 6	5,2	6,5	8,6	6,0	7,2	10,1
7-8	5,1	6,1	7,3	5,2	6,2	8,1
9-10	5,1	6,2	7,8	5,0	6,4	8,0
11-12	5,3	6,2	7,5	5,1	6,4	7,9
13-14	5,2	6,2	8,0	5,0	6,2	7,2
15-16	5,1	6,3	8,0	5,0	6,3	7,6
17-18	5,0	6,1	7,5	5,0	6,3	7,5
19-20	5,2	6,4	7,8	5,0	6,5	7,8
21-22	4,5	6,0	7,1	5,4	6,2	7,7
23-24	5,0	6,0	7,5	5,2	6,4	7,5
≥ 25	5,1	6,2	7,6	5,0	6,3	7,8

Table 13. Mean length of central apnea during active sleep, sec. (Kelmanson IA, 1995)

Age, weeks	Centiles for distribution					
	Girls			Boys		
	10	50	90	10	50	90
≤ 6	5,4	5,9	6,5	5,6	6,1	6,7
7-8	5,1	5,7	6,2	5,3	5,7	6,3
9-10	5,1	5,6	6,1	5,2	5,6	6,2
11-12	5,2	5,6	6,0	5,2	5,7	6,2
13-14	5,2	5,6	6,0	5,2	5,6	6,1
15-16	5,1	5,6	6,1	5,0	5,6	6,2
17-18	5,2	5,7	6,1	5,1	5,5	6,1
19-20	5,2	5,7	6,3	5,3	5,8	6,3
21-22	5,3	5,6	6,2	5,0	5,6	6,4
23-24	5,1	5,7	6,2	5,2	5,7	6,4
≥ 25	5,3	5,8	6,4	5,2	5,8	6,5

Table 14. Percent of time falling on central apneas during quiet sleep (Kelmanson IA, 1995)

Age, weeks	Centiles for distribution					
	Girls			Boys		
	10	50	90	10	50	90
≤ 6	0	0,3	1,1	0,2	0,7	1,9
7-8	0	0,4	1,2	0	0,4	1,2
9-10	0,1	0,5	1,4	0,1	0,4	1,1
11-12	0,1	0,5	1,2	0,1	0,5	1,3
13-14	0,1	0,4	0,9	0,1	0,4	0,9
15-16	0,1	0,4	0,9	0	0,4	1,0
17-18	0,1	0,3	1,0	0	0,3	0,9
19-20	0,1	0,4	0,6	0	0,3	1,0
21-22	0	0,2	0,7	0	0,3	0,9
23-24	0,1	0,4	0,8	0	0,3	0,7
≥ 25	0,1	0,3	0,7	0	0,3	0,8

Table 15. Percent of time falling on central apneas during active sleep (Kelmanson IA, 1995)

Age, weeks	Centiles for distribution					
	Girls			Boys		
	10	50	90	10	50	90
≤ 6	0,3	1,4	4,0	0,6	1,6	5,7
7-8	0,4	1,2	2,5	0,4	0,9	2,8
9-10	0,4	1,0	3,0	0,3	1,0	2,2
11-12	0,4	1,0	2,5	0,3	1,0	2,1
13-14	0,4	1,1	2,1	0,4	0,8	1,7
15-16	0,4	1,0	1,8	0,3	0,9	2,1
17-18	0,3	0,9	2,0	0,2	0,8	1,7
19-20	0,6	0,9	1,8	0,4	0,9	1,6
21-22	0,4	0,9	2,0	0,2	0,8	2,4
23-24	0,4	1,2	2,1	0,4	0,9	2,2
≥ 25	0,4	1,0	2,1	0,3	0,8	1,6

Normally, CA are not accompanied by the marked reduction in blood oxygen saturation, nor by the noticeable bradycardia or other functional disorders. Length of the CA seldom exceeds 20 seconds, even in the youngest infants and children. One of the indirect signs of the maturation of respiratory mechanisms is the reduction in the number of central respiratory pauses during sleep, and this is particularly evident during first 3 months of infant's life (Kelmanson IA, 1995; 1997).

There is a specific form of respiration known as *periodic breathing (PB)*. PB is characterized by three or more respiratory pauses lasting 3 seconds or more, separated by the episodes of normal respiratory activity lasting 20 seconds or less (Figure 7). PB, as a specific form of intermittent CA in the young infants, is not uncommon in this age group (Table 16). However, when total duration of PB exceeds 5 % of the total sleep time, one can speak about pathological periodic breathing. In the full-term newborn infants, PB is mainly observed during active sleep and becomes rare after 3-4 months. Therefore, respiratory pauses of the central origin during sleep and irregular respiration during sleep should be considered as the transitory phenomenon, reflecting immaturity of the central controlling mechanisms. On the contrary, complete absence of the CA during sleep in these infants should be regarded as the warning sign and should be considered cautiously for possible serious neurological abnormalities.

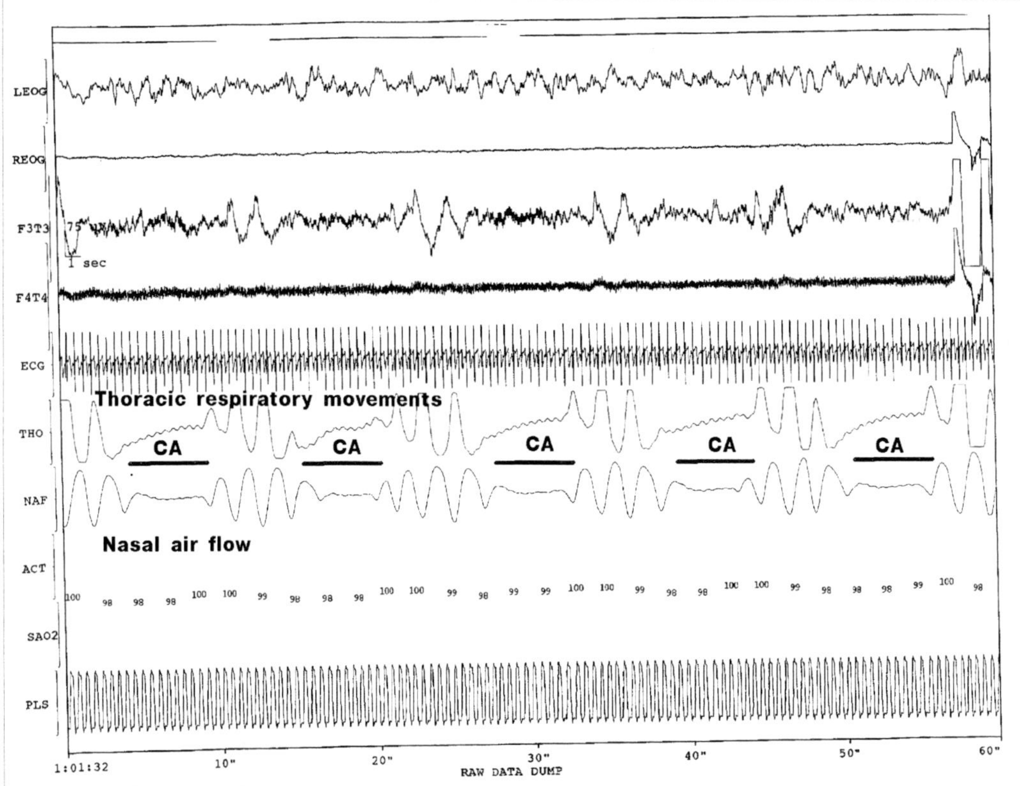

Figure 7. Polysomnographic picture of periodic breathing in the infant

Unlike CA, OA and MA are normally less common in the infants and young children (Tables 17-28). They are practically absent during quiet sleep, and normally are not accompanied by essential reduction in oxygen saturation, marked bradycardia; seldom they last more than 20 seconds, and their frequency reaches 3 apneas per 1 hour of sleep only in the youngest infants; afterwards, it does not exceed 1 apnea per 1-2 hours.

Table 16. Percent of time falling on periodic breathing during sleep (Kelmanson IA, 1995)

Age, weeks	Centiles for distribution					
	Girls			Boys		
	10	50	90	10	50	90
≤ 6	0	1,2	7,2	0	1,3	8,8
7-8	0	0,7	5,3	0	0,4	2,4
9-10	0	0,6	3,9	0	0,4	2,7
11-12	0	0,4	2,6	0	0,3	1,5
13-14	0	0,4	2,0	0	0,3	1,2
15-16	0	0,3	1,4	0	0,2	1,2
17-18	0	0,3	1,9	0	0,2	1,3
19-20	0	0,3	1,1	0	0,3	0,8
21-22	0	0,4	1,3	0	0,3	1,6
23-24	0	0,3	2,0	0	0,2	1,2
≥ 25	0	0,2	1,3	0	0,2	0,8

Table 17. Number of obstructive apneas per 1 hour of total sleep time (Kelmanson IA, 1995)

Age, weeks	Centiles for distribution					
	Girls			Boys		
	10	50	90	10	50	90
≤ 6	0	0,4	3,0	0	0,4	1,6
7-8	0	0,1	0,9	0	0,1	0,9
9-10	0	0	0,7	0	0,1	0,8
11-12	0	0,1	0,5	0	0,1	0,6
13-14	0	0	0,7	0	0	0,8
15-16	0	0,1	0,8	0	0,1	0,8
17-18	0	0,1	0,9	0	0	0,6
19-20	0	0,1	0,6	0	0,1	1,4
21-22	0	0	0,7	0	0,1	0,9
23-24	0	0,1	0,7	0	0,1	1,1
≥ 25	0	0,1	0,7	0	0,1	0,9

Table 18. Number of obstructive apneas per 1 hour of active sleep (Kelmanson IA, 1995)

Age, weeks	Centiles for distribution					
	Girls			Boys		
	10	50	90	10	50	90
≤ 6	0	0,6	3,6	0	0,4	1,9
7-8	0	0,2	1,0	0	0,1	0,9
9-10	0	0	1,0	0	0,2	1,2
11-12	0	0	0,8	0	0,2	0,9
13-14	0	0	0,9	0	0	1,2
15-16	0	0,2	1,2	0	0,1	1,2
17-18	0	0	1,4	0	0	0,9
19-20	0	0	1,1	0	0,2	2,2
21-22	0	0	1,3	0	0,2	1,5
23-24	0	0,1	1,2	0	0,3	2,0
≥ 25	0	0,2	1,1	0	0,2	1,8

Table 19. Maximal length of obstructive apnea during sleep, sec. (Kelmanson IA, 1995)

Age, weeks	Centiles for distribution					
	Girls			Boys		
	10	50	90	10	50	90
≤ 6	4,7	8,8	17,5	4,0	7,5	13,0
7-8	3,0	5,5	10,0	3,3	6,5	11,4
9-10	3,5	5,0	9,0	3,5	5,5	8,9
11-12	3,0	5,0	8,0	3,5	5,5	8,9
13-14	3,5	5,0	8,5	3,2	5,5	8,5
15-16	3,5	5,5	8,5	3,5	5,5	9,0
17-18	3,0	5,0	7,6	3,5	5,5	10,4
19-20	3,0	5,3	11,5	3,4	6,5	14,1
21-22	3,1	4,5	12,7	3,3	5,8	8,6
23-24	4,0	4,8	8,0	4,0	5,0	13,6
≥ 25	3,5	5,5	11,7	3,5	6,0	10,8

Table 20. Maximal length of obstructive apnea during active sleep, sec. (Kelmanson IA, 1995)

Age, weeks	Centiles for distribution					
	Girls			Boys		
	10	50	90	10	50	90
≤ 6	4,7	8,3	15,8	4,0	7,5	13,0
7-8	3,0	5,5	10,0	3,3	6,3	11,5
9-10	3,5	5,0	9,0	3,5	5,5	9,0
11-12	3,0	5,0	8,0	3,5	5,5	8,5
13-14	3,5	5,0	8,5	3,0	5,5	8,5
15-16	3,5	5,5	8,0	3,4	5,5	9,0
17-18	3,0	5,0	7,5	3,5	5,0	10,4
19-20	3,0	5,0	10,8	3,3	6,0	14,3
21-22	3,1	4,5	9,0	4,0	5,5	8,5
23-24	4,0	4,5	8,0	3,6	5,0	13,7
≥ 25	3,5	5,0	11,9	3,5	6,0	11,1

Table 21. Mean length of obstructive apnea during active sleep, sec. (Kelmanson IA, 1995)

Age, weeks	Centiles for distribution					
	Girls			Boys		
	10	50	90	10	50	90
≤ 6	4,0	5,7	7,8	3,7	5,5	7,3
7-8	3,0	4,5	6,5	3,3	4,9	7,9
9-10	3,5	4,5	7,0	3,5	4,5	7,0
11-12	3,0	4,5	6,5	3,5	4,5	6,0
13-14	3,5	4,2	6,0	3,0	4,5	6,1
15-16	3,5	4,5	6,0	3,2	4,7	6,5
17-18	3,0	4,1	5,8	3,5	4,5	6,7
19-20	3,0	4,0	5,5	3,1	4,5	6,6
21-22	3,0	3,8	5,2	3,5	4,8	6,5
23-24	3,5	4,0	6,8	3,3	4,5	6,8
≥ 25	3,2	4,5	7,0	3,3	4,6	6,8

Table 22. Percent of time falling on obstructive apneas during active sleep (Kelmanson IA, 1995)

Age, weeks	Centiles for distribution					
	Girls			Boys		
	10	50	90	10	50	90
≤ 6	0	0,1	0,5	0	0,1	0,3
7-8	0	0	0,1	0	0	0,2
9-10	0	0	0,1	0	0	0,2
11-12	0	0	0,1	0	0	0,1
13-14	0	0	0,1	0	0	0,2
15-16	0	0	0,2	0	0	0,1
17-18	0	0	0,2	0	0	0,1
19-20	0	0	0,1	0	0	0,3
21-22	0	0	0,1	0	0	0,2
23-24	0	0	0,1	0	0	0,3
≥ 25	0	0	0,2	0	0	0,3

Table 23. Number of mixed apneas per 1 hour of total sleep time (Kelmanson IA, 1995)

Age, weeks	Centiles for distribution					
	Girls			Boys		
	10	50	90	10	50	90
≤ 6	0	0,2	1,7	0	0,3	2,0
7-8	0	0	0,3	0	0	0,4
9-10	0	0	0,3	0	0	0,3
11-12	0	0	0,3	0	0	0,2
13-14	0	0	0,3	0	0	0,3
15-16	0	0	0,2	0	0	0,3
17-18	0	0	0,2	0	0	0,1
19-20	0	0	0,1	0	0	0,3
21-22	0	0	0,2	0	0	0,1
23-24	0	0	0,1	0	0	0,3
≥ 25	0	0	0,2	0	0	0,1

OA index exceeding 1 event per 1 hour of the total sleep time should be considered as abnormal, irrespective of the length of apnea.

In the older children and in the adults, most apneas during sleep are OA and MA, while CA are very uncommon, and, if observed, should be considered as the sign of major impairments of the CNS

Table 24. Number of mixed apneas per 1 hour of active sleep (Kelmanson IA, 1995)

Age, weeks	Centiles for distribution					
	Girls			Boys		
	10	50	90	10	50	90
≤ 6	0	0,2	1,8	0	0,4	2,2
7-8	0	0	0,4	0	0	0,4
9-10	0	0	0,4	0	0	0,4
11-12	0	0	0,4	0	0	0,3
13-14	0	0	0,4	0	0	0,4
15-16	0	0	0,3	0	0	0,4
17-18	0	0	0,3	0	0	0,3
19-20	0	0	0,3	0	0	0,5
21-22	0	0	0,3	0	0	0,2
23-24	0	0	0,2	0	0	0,5
≥ 25	0	0	0,3	0	0	0,3

Table 25. Maximal length of mixed apnea during sleep, sec. (Kelmanson IA, 1995)

Age, weeks	Centiles for distribution					
	Girls			Boys		
	10	50	90	10	50	90
≤ 6	6,4	10,3	19,3	6,0	11,8	19,1
7-8	5,2	7,8	12,5	5,0	6,5	11,2
9-10	4,5	7,0	12,1	5,0	7,5	11,1
11-12	5,0	7,0	10,0	5,0	7,0	10,6
13-14	4,5	6,0	11,6	5,0	6,0	9,1
15-16	4,5	6,0	9,6	5,0	6,5	10,5
17-18	4,8	6,3	10,0	4,5	6,5	9,5
19-20	4,5	6,5	12,2	4,8	6,5	14,0
21-22	4,7	6,0	13,7	5,0	6,5	11,0
23-24	4,5	6,0	13,0	4,5	7,5	11,0
≥ 25	4,9	7,5	11,1	4,5	6,8	13,0

The specific respiratory phenomena, often observed in the infants and young children, are the *sighs*. A sigh is defined as an isolated breath with an increased tidal volume that is seen during sleep or wakefulness. Sighs may occur spontaneously or be induced by lung inflation or airway occlusion (Bartlett D, 1971; Thach BT, Taeusch W, 1976). Sighs then result from a vagally-induced inspiration-augmented reflex and rapidly adapting pulmonary mechano-receptors. Sighs are associated with increases in pulmonary compliance and functional residual capacity, and the opening of collapsed alveoli (Perez-Padilla R et al.,

1983). In sleeping infants, sighs are associated with sleep apneas (Alvarez JA et al., 1993; Davies GM, Moscato J, 1994; Weintraub Z et al., 1994), and fluctuations in heart rates or blood oxygenation (Eiselt M et al., 1992). Polysomnographically, a sigh is defined as a brisk and isolated increase in thoracoabdominal excursion with an amplitude at least twice superior to that measured during the 10 seconds that preceded the event. Nonspecific body movements are distinguished from sighs by the presence of movement artifacts on the thermistors and EEG tracings. When more than one sharp increase in respiratory and cardiac amplitude occurs in a row, the event should also be scored as a movement.

Table 26. Maximal length of mixed apnea during active sleep, sec. (Kelmanson IA, 1995)

	Centiles for distribution					
Age, weeks	Girls			Boys		
	10	50	90	10	50	90
≤ 6	6,4	9,3	19,3	6,0	11,0	19,1
7-8	5,1	7,5	12,5	5,0	6,5	11,4
9-10	4,5	7,0	10,8	5,0	7,5	11,3
11-12	5,0	7,0	10,2	5,0	7,0	10,5
13-14	4,5	6,0	10,2	5,0	6,0	9,1
15-16	4,5	6,0	9,6	5,0	6,5	10,5
17-18	5,0	6,5	10,3	4,5	6,5	9,6
19-20	4,5	6,5	12,4	4,5	6,5	14,1
21-22	4,6	6,5	14,3	5,0	6,5	12,3
23-24	4,5	6,0	12,2	5,0	7,5	11,4
≥ 25	4,5	7,3	10,8	4,5	7,0	13,0

Table 27. Mean length of mixed apnea during active sleep, sec. (Kelmanson IA, 1995)

	Centiles for distribution					
Age, weeks	Girls			Boys		
	10	50	90	10	50	90
≤ 6	5,9	8,1	9,7	5,7	8,4	12,6
7-8	5,0	7,0	11,9	4,8	6,5	10,1
9-10	4,5	6,5	9,1	5,0	7,0	9,4
11-12	5,0	6,5	8,1	5,0	6,5	9,7
13-14	4,5	5,9	8,6	5,0	5,9	8,0
15-16	4,5	6,0	8,6	4,8	6,0	8,1
17-18	4,8	6,0	9,1	4,5	6,5	9,6
19-20	4,5	6,3	10,3	4,5	6,0	9,1
21-22	4,6	6,0	14,3	5,0	6,0	9,5
23-24	4,5	5,9	11,2	4,8	6,4	10,0
≥ 25	4,5	6,9	9,4	4,5	6,5	9,2

Table 28. Percent of time falling on mixed apneas during active sleep (Kelmanson IA, 1995)

Age, weeks	Centiles for distribution					
	Girls			Boys		
	10	50	90	10	50	90
≤ 6	0	0	0,4	0	0,1	0,5
7-8	0	0	0,1	0	0	0,1
9-10	0	0	0,1	0	0	0,1
11-12	0	0	0,1	0	0	0,1
13-14	0	0	0,1	0	0	0,1
15-16	0	0	0,1	0	0	0,1
17-18	0	0	0,1	0	0	0,1
19-20	0	0	0,1	0	0	0,1
21-22	0	0	0,1	0	0	0,1
23-24	0	0	0,1	0	0	0,1
≥ 25	0	0	0,1	0	0	0,1

Sighs may be classified in four categories depending on their association with apneas. *Isolated sighs*are scored when they were not preceded and not followed by an apnea; *pre-apneic sighs* are immediately followed by an apnea; *post-apneic sighs* are immediately preceded by an apnea; *intra-apneic sighs* are immediately preceded and followed by an apnea.

Pre-apnea sighsare seen most frequently in the infants, followed by isolated sighs, post-apnea sighs, and intra-apnea sighs. Sighs may be found in all sleep states but are more frequently seen in NREM/quiet sleep, than in REM/active sleep (Alvarez JE et al., 1993; Kelmanson IA et al., 2003). The apneas that follow the sighs may be initiated by a rapid decline in carotid body chemo-receptor afferent discharge, due to the sigh-induced increase in arterial oxygen concentration and decrease in blood carbon dioxide (Perez-Padilla R et al., 1983; Fleming P et al., 1984). The association between sighs and apneas could however result from the complex interactions of factors, including peripheral chemo-receptor sensitivity, the degree of maturation and behavioral state, or central control mechanisms, including forebrain.

Hypoxia promotes more frequent occurrence of sighs in infants and young children. In infants born full-term, sighs during quiet sleep initially lead to increased heart rate and then to its deceleration. During active sleep, sighs may only lead to increased heart rate. In the infants born pre-term at 35 week of gestation and later, in either sleep state sighs are accompanied by accelerated heart rate; in deeply premature infants born at gestational age 31-34 wk, sighs do not lead to changes in heart rate (Eiselt M et al., 1992).

Sleep and its certain stages are known for changes in *cardiac rhythm and major hemodynamic characteristics*. During sleep stage 1 and at transition to stage 2, mean heart rate and heart rate variability decrease, and quiet/NREM sleep is known for the minimal daily values of the heart rate and arterial blood pressure. During SWS, heart rate is markedly reduced and heart rate variability is decreased. Cardiac rhythm during REM/active sleep becomes extremely unstable: mean heart rate is higher than during NREM/quiet sleep, and heart rate variability is markedly increased (Kelmanson IA, 1995; 1997) (Tables 29-33).

Table 29. Mean heart rate in the infants during active sleep (Kelmanson IA, 1995)

Age, weeks	Centiles for distribution					
	Girls			Boys		
	10	50	90	10	50	90
≤6	127	146	155	129	138	151
7-8	123	136	150	125	139	151
9-10	123	133	142	122	132	145
11-12	121	131	142	121	130	140
13-14	118	130	140	118	129	139
15-16	118	129	141	117	128	138
17-18	116	129	138	116	127	138
19-20	118	126	143	111	123	141
21-22	113	126	137	114	124	135
23-24	108	122	140	110	120	133
≥25	111	123	136	108	121	134

Table 30. Mean heart rate in the infants during quiet sleep (Kelmanson IA, 1995)

Age, weeks	Centiles for distribution					
	Girls			Boys		
	10	50	90	10	50	90
≤6	120	139	152	122	134	146
7-8	117	129	144	117	133	147
9-10	115	127	138	116	126	139
11-12	114	126	137	115	125	136
13-14	111	124	133	112	121	135
15-16	109	123	133	109	121	132
17-18	110	120	131	110	120	131
19-20	107	118	135	104	118	133
21-22	108	120	129	107	117	128
23-24	102	114	133	102	112	128
≥25	103	117	127	100	114	127

The average arterial BP decreases by 5-9 % during sleep stages 1 and 2 and by 8-14% during stages 3 and 4, compared with a wake state. During active sleep, marked fluctuations in BP are found, but generally there is a trend towards its increase by 3-6 % as compared with quiet sleep (Coote JH, 1982). Cyclic fluctuations of blood pressure with amplitude of 15-20 mm Hg. and periodicity of 20-30 seconds may be recorded (Martin RJ, 1987). This periodicity coincides with fluctuations of the air flow speed in the lungs and with fluctuations of the heart rate.

Table 31. Heart rate variability in the infants during active sleep9 (Kelmanson IA, 1995)

	Centiles for distribution					
Age, weeks	Girls			Boys		
	10	50	90	10	50	90
≤ 6	5,8	7,8	10,0	6,2	8,5	11,0
7-8	5,9	8,0	13,4	6,7	8,5	11,0
9-10	6,4	8,8	11,7	6,5	8,9	12,6
11-12	6,8	8,9	11,7	6,6	8,7	12,7
13-14	6,5	8,7	12,9	6,9	9,3	13,6
15-16	6,5	9,0	13,4	6,9	8,5	12,4
17-18	6,2	9,0	13,1	6,9	8,8	12,1
19-20	7,0	8,5	11,9	6,7	8,4	13,9
21-22	6,3	8,8	11,3	7,1	8,6	12,4
23-24	6,5	10,0	13,6	7,2	9,4	14,0
≥ 25	6,6	9,1	14,7	7,2	9,7	14,0

[1] Heart rate variability is calculated as standard deviation of the heart rate devided by the mean heart rate and multiplied by 100 %

Table 32. Heart rate variability in the infants during quiet sleep (Kelmanson IA, 1995)

	Centiles for distribution					
Age, weeks	Girls			Boys		
	10	50	90	10	50	90
≤ 6	2,5	3,9	5,5	3,0	4,1	5,9
7-8	2,8	4,0	6,3	3,3	4,8	7,2
9-10	3,1	4,5	6,4	3,2	4,2	6,2
11-12	3,2	4,4	6,6	3,2	4,5	6,3
13-14	3,0	4,5	6,7	3,3	4,6	6,8
15-16	3,4	4,6	7,0	3,2	4,4	6,6
17-18	3,2	4,5	8,1	3,4	4,4	6,2
19-20	3,3	4,7	7,0	3,9	5,0	7,5
21-22	3,3	4,7	6,4	3,4	4,5	7,1
23-24	3,7	5,0	7,3	3,8	5,3	7,9
≥ 25	3,6	5,2	7,5	3,9	5,3	7,1

Presumably, this may result from the cyclic changes in the activity of the cerebral structures responsible for the regulation of hemodynamics and from the partial obstruction of the UAW that normally occurs during sleep. Cardiac output decreases during transition from wake to active sleep. During quiet sleep, it is reduced by 7-12 %, partly due to the decreased heart rate. Active sleep is characterized by maximal (up to 25 %) decrease in cardiac output, mostly due to the reduced stroke volume (Miller JC, Horvath SM, 1976). Pulmonary artery pressure changes during sleep, and, as a rule, both systolic and diastolic pressure are increased, although the values are usually within the normal limits. Probably, the major cause of increased pulmonary artery pressure during sleep is the reduced blood oxygen tension. Potential involvement of the reflexes connected with pulmonary artery stretching is also discussed.

Releasing of arachidonic acid and its metabolites, reduction in concentration of active vasodilators, as well as the reduction in activity of beta-adrenergic receptors may also contribute (Martin RJ, 1987).

Compared with the wake state, heart rate during NREM/quiet sleep is reduced by 5-8%. This is caused by a reduction in sympathetic tone in this sleep stage. Cardiac respiratory arrhythmia is mostly pronounced during NREM/quiet sleep, which is partly connected with increased duration of inspiration. In the sleep stages 3 and 4, there is maximal link between the rhythm of respiration and the cardiac rhythm, and this respiratory modulation of the cardiac rhythm is expressed much more than in a wake state. This pronounced synchronization can be used for the identification of the SWS (Harper RM et al., 1987). Association between the sleep stages and cardiac respiratory arrhythmia depends on infant's and child age: at the early stages of postnatal development, amplitude of heart rate fluctuations during sleep slightly decreases, and then starts to increase with advancing age (Harper RM et al., 1978b), and this does not depend on absolute values the heart rate (Kluge KA et al., 1988).

During active/REM-sleep, heart rate reaches the values characteristic for a state of wakefulness. This may be due to increased sympathetic tone during that sleep stage, which, sometimes, may even exceed the values noticeable in the wake state. Increased sympathetic tone is observed in the form of short-term episodes, bursts that often, although not always, occur simultaneously with the episodes of rapid eye movements (Wallin BG, 1991). Increased power of non-respiratory periodic components of cardiac rhythm variation is usually registered during these episodes (Coccagna G et al., 1971). Compared with quite/NREM sleep and wake state, active/REM sleep is characterized by augmented oculo-cardiac reflex in infants and children (Ramet J et al., 1991). Importantly, reaction to vagal stimulations during active/REM-sleep in the form of bradycardia and/or other cardiac arrhythmias is most clearly expressed in infants and young children, especially born with low birth weight and/or pre-term. With advancing age and maturation these vagal-induced reactions become less prominent (Ramet J, 1989). The periods of full dissociation between cardiac and respiratory rhythm with intermittent episodes of their close association may be observed during active/REM sleep (Bond WC et al., 1973). Sinus arrest lasting several seconds may occur during active/REM sleep. This may not be accompanied by the changes in respiratory activity and reduced oxygen saturation; these episodes may not induce arousal reaction or full awakening, thus making the situation particularly threatening (Guilleminault C et al., 1984b).

Chapter VII

Arousals

Arousal is a transition to more superficial sleep stages or full awakening. This reaction is an important protective mechanism against various life-threatening circumstances. Arousal can manifest as a short-term episodes of increased muscular tone, which is defined by EMG, as a transition from one sleep stage to another, as the short periods of a EEG desynchronization, as well as the full awakening (Phillipson EA, Sullivan CE, 1978). If the healthy adults are exposed to acoustic stimuli aimed to awoke arousals, the most common registered form of the given reaction would be an increase in EMG activity (24 % cases), followed in the order of reduced frequency by body movements (19 %), transition to more superficial sleep stages (19 %), increased EEG frequency (17 %), appearance of the alpha waves on EEG (13 %) and awakening (2 %) (Badia P et al., 1985). Apart from the specified changes, arousals may be accompanied by transitory increase in BP, increased heart rate and respiratory rate, which is the evidence of increased sympathetic nervous system tone. The manifestation of activation is disproportionate to the effort required for transition from a state of sleep to wakefulness, and the sympathetic activation accompanying arousals is directed at organism's mobilization before potential menace that was a cause of arousal (Horner RL, 1996).

From the point of view of neuro-anatomy, arousal is not connected with any specific brain nucleus; however it is possible to specify some brain structures involved in this reaction. These are pons, posterior parts of hypothalamus, sub-thalamic departments, basal parts of the forebrain (Jones BE, 1989). During arousals, the activating reticular formation of the brain receives collateral impulses from visceral, somatic and other sensory systems, then transmitting the signals to the forebrain, which is manifested by cortical activation (Jones BE, 1989).

Standard methods for scoring arousability in infants and young children are still lacking (Loredo JS et al., 1999). The literature is scarce, includes a wide range of terminologies and leaves the problem unsolved. The American Sleep Disorders Association (ASDA) proposed definitions of arousals for adults (Association ASD, 1992). According to ASDA definition, during NREM sleep, arousal can be scored, if abrupt shift in EEG frequency occurs, including appearance of the theta- and alpha-waves, as well as episodes of EEG-activity with the frequency exceeding 16 Hz other than sleep spindles. Duration of these EEG changes should be over at least 3 seconds, and the subsequent continuous sleep should last not less than 10 seconds. During REM sleep, arousal can be scored when the above changes are observed, accompanied by the increase in the EMG amplitude (Association ASD, 1992). Increased EMG amplitude, either during NREM or REM sleep, if not accompanied by EEG changes, should not be scored as arousal. In the newborn and the young infant, the well-known spontaneous variability in respiratory, body movements and heart rate, significantly complicate the evaluation of arousals. Hence, the use of the ASDA definition appears not be appropriate for infants. The International Pediatric Wake-up Club was set up with the aim to define a method for the scoring of arousals in infants, based on the analysis of PSG sleep-wake recordings (Scoring of arousals in healthy term infants (between the ages of 1 and 6 months), 2005).

A *sub-cortical activation* and *cortical* arousal are differentiated (Figure 8; Figure 9). Baseline sleep state should be established during 30 s epoch. At least 10 s of uninterrupted state is required between cortical arousals. The duration of the reference period should be described.

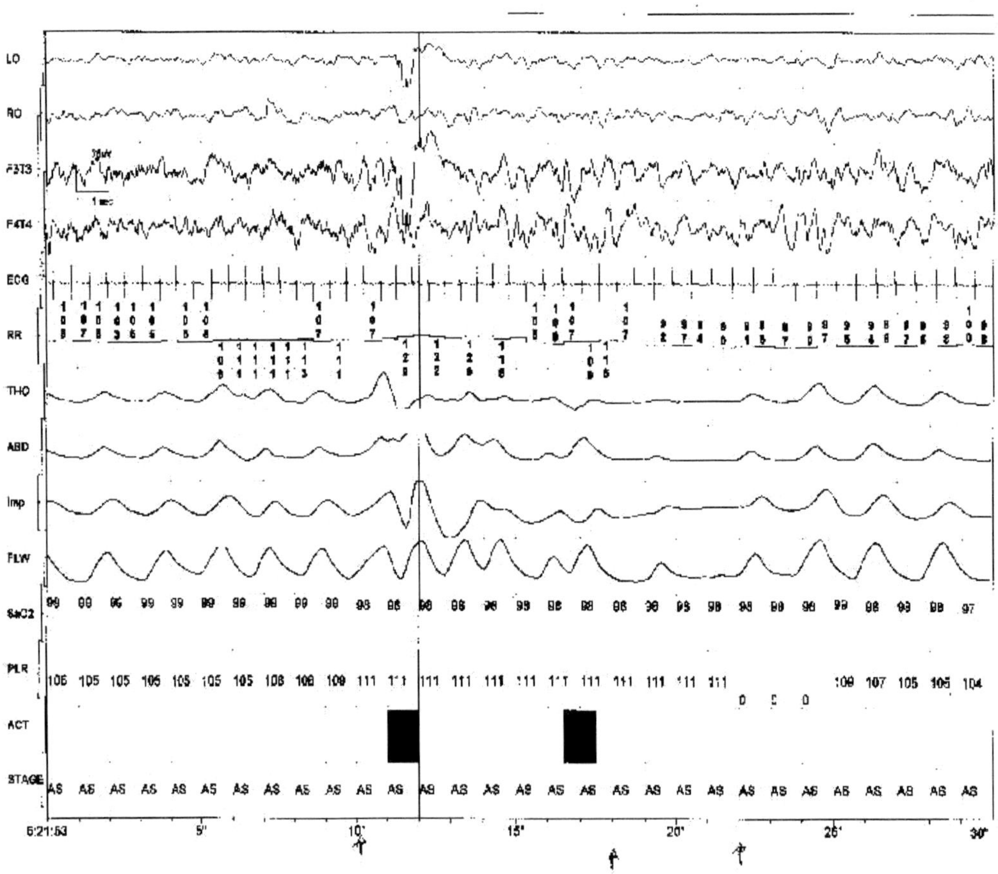

Figure 8. Polysomnographic picture of sub-cortical activation in the infant

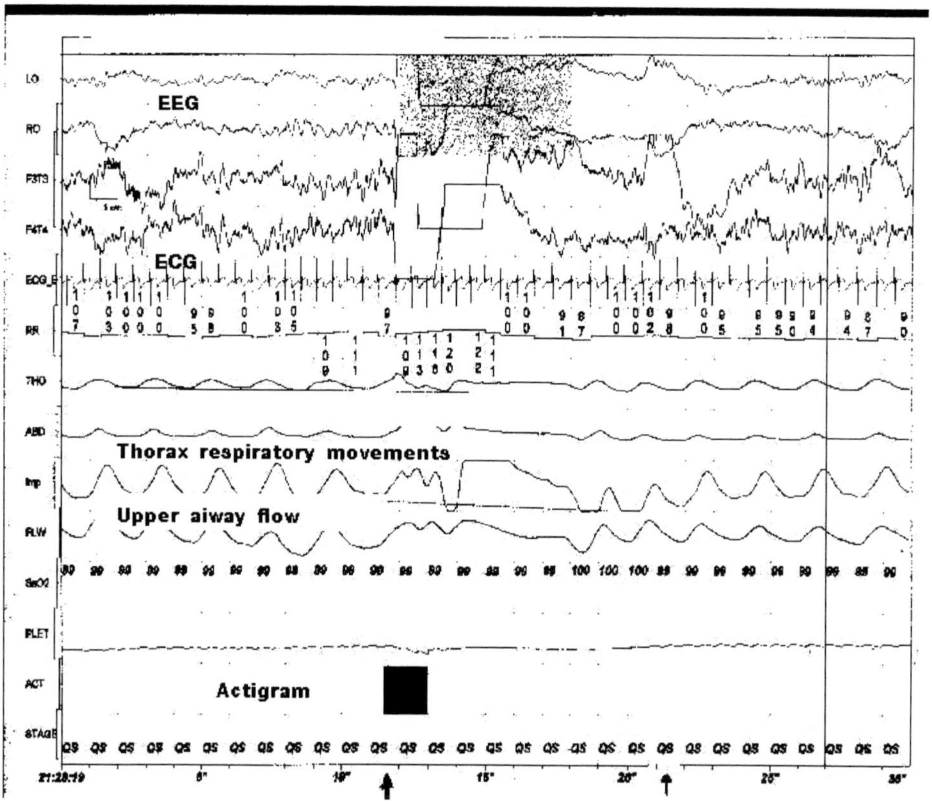

Figure 9. Polysomnographic picture of cortical arousal in the infant

A sub-cortical activation is scored if no change in EEG is seen, but at least two of the following changes are apparent during at least 3 sec, during either NREM or REM sleep:

a) In NREM sleep:

- A gross body movement detected by movement sensors or seen by movement artefacts in the somatic channels (ECG, EEG, respiratory parameters) or by direct observation. Such movements may include startles.
- Changes in heart rate (at least 10% faster than baseline values).
- Changes in breathing pattern (frequency and/or amplitude). These changes may include a single augmented breaths.

b) In REM sleep:

- A gross body movement detected by movement sensors or seen as movement artefacts in the somatic channels (ECG, EEG, respiratory parameters) or by observation.
- Changes in heart rate (at least 10% faster than baseline values).
- An increase in chin EMG amplitude, unless associated with sucking, that was not previously present.

A cortical arousal is scored using the above criteria, with in addition, an abrupt change in EEG background frequency (of at least 1 Hz), for a minimum of 3 seconds.

A decrease in EEG amplitude can help identify cortical arousals.

Cortical arousals and sub-cortical activations are subdivided in two major categories: *spontaneous* (no stimulus can be identified) and *induced* (following an identifiable stimulus). Spontaneous cortical arousals or sub-cortical activation occur for no detected cause. Some may be associated with changes in vigilance, such as changes in sleep states, often accompanied by body movements. These arousals or sub-cortical activations are classified as «spontaneous, associated with sleep-state changes».

Induced cortical arousals and sub-cortical activations are classified according to the suspected cause:

1. Internal Stimuli
   - Respiratory related cortical arousals or sub-cortical activations, associated with apnea (central, mixed, obstructive), hypopneas, hypoxia/hypercapnia, increased respiratory efforts.
   - Other system related, e.g. cardiac (e.g., bradycardia), muscular, digestive (e.g., reflux, colics).
2. External Stimuli

   Cortical arousals or sub-cortical activations that follow changes in the environment of the child (such as auditory, visual, temperature, chemical or tactile stimuli).

   The following information can contribute to a complete description of the observed changes:

   a) The preceding sleep state. Changes on the PSG are evaluated by comparison with a pre-arousal reference period that immediately precedes the arousal or the sub-cortical activation. The duration of the reference period should be chosen according to the aim of the study and should be specified.
   b) The complete sequence of recorded changes. The start of the arousal to the return to baseline for all the parameters is considered a complete sequence of the arousal.
   c) The following sleep state. A change in state, be it awake or another sleep state, is not an arousal.

The propensity to arouse from sleep can be evaluated by exposing the infant to awakening challenges. An arousal threshold is determined by measuring the intensity of the stimulus needed to induce an arousal. The time spent between the start of the stimulus and the arousal reaction can also be evaluated. The study of the arousal threshold requires that the subject investigated spends a long enough period asleep to exhibit all classic sleep stages and to permit continuous PSG recordings of EEG, ECG, air flux, thoracic and abdominal movements in order to monitor EEG as well as non-EEG changes. Stimuli used in the evaluation of arousability should be non-invasive, easily quantifiable and induce specific as well as reproducible responses.

It is not clear from the literature how a threshold should be determined. It can be scored following a single reaction to an increasing - or decreasing - series of stimulations. It can also require several intensities of the stimuli administered both above and under the level of reaction, so as to define supra- and infraliminar intensities of the challenge (Read PA et al., 1998). No consensus exists on the type of challenge that best fits the determination of the thresholds. Noise, gases, light, nociceptive, mechanical, chemical or temperature stimuli have all been used. Individual characteristics as well as experimental conditions greatly modify the subjects' arousal responses.

Auditory stimulation is among the widely used approaches. No standard method exists for the administration of the auditory stimuli. Auditory arousal challenges have been performed either with unfamiliar stimuli such as white noise (Franco P et al., 2001), 1-kHz pure tone (Thomas DA et al., 1996), or familiar noises (telephone ring, train, etc.) (Franco P et al., 2000a). Various strategies have been used to administer auditory stimulations. The signal generators may be located in the midline of the cot or near the ears of the infants. The initial signals were either constant stimuli with fixed tonus or with increasing tone intensity . The lapse between stimulations also varied from some milliseconds to a minute. Apart from the methodological differences, this method has been well studied. The reproducibility of the method was demonstrated. There is a significant night-to-night correlation in auditory threshold during NREM sleep and REM sleep, at least in the adult population.

Light stimulations appear to be a weak arousal stimulus in infants (Davidson Ward SL et al., 1992a; Busby KA et al., 1994). Only 47% of infants had behavioral arousals with photo-stimulator in NREM sleep (Davidson Ward SL et al., 1992a).

Controlled pulse air jet evokes trigeminal stimulation by applying a jet of air to the nostrils in varying intensities (Read PA et al., 1998).

Vibro-tactile stimuli excite muscle receptors and/or pacenian corpuscules by mechanical stimulations at different frequencies and intensities on the arm (Newman NM et al., 1989).

Head-up tilting test. Infants were tilted 60° head-up and arousal responses were observed (Galland BC et al., 1998). The postural change inducing a fall in blood pressure stimulates the orthostasis arterial baroreceptors.

Arousal stimuli related to respiration. Respiratory arousal is defined as arousal from sleep associated with progressive increases in stimuli related to respiration (hypoxia, hypercapnia, and respiratory effort) (Berry RB, Gleeson K, 1997).

The arousal response *to hypoxic or hypercapnic challenges* is influenced by sleep stages, the duration of the exposition, the depth of hypoxia, or the method of delivery of the gas mixture (Berry RB, Gleeson K, 1997). Hypoxic challenges show marked interindividual and interspecies variations in arousal thresholds. Hypoxemia is considered to be a poor arousal stimulus in infants, while hypercapnia is a potent arousal stimulus. In real life conditions however, hypoxemia is likely associated with a mild and transient increase in end tidal CO_2.

Arousal from *upper airway narrowing or occlusion* appears to involve more stimuli that the arterial blood gas changes alone, as arousal can occur at the termination of apneas or hypopneas that are too brief for substantial asphyxia blood gas changes to develop (Berry RB, Gleeson K, 1997).

A spontaneous arousal can occur during an arousal challenge. It may then be difficult to evaluate whether the infant woke up spontaneously or responded to the challenge.

Various endogeneous stimuli induce adaptive responses in the form of arousals. These factors include prolonged apneas (Rees K et al., 1995), acid esophageal reflux reaching the proximal esophagus (Kahn A et al., 1991b), leg movements (Winkelman JW, 1999), pain (Oberlander TF et al., 1999), tracheobronchial irritation (Jeffery HE et al., 1999), changes in body temperature (Karacan I et al., 1968) or in blood pressure (Trelease RB et al., 1985). The awakening effects of endogeneous stimuli depend on multiple conditions. The effects of laryngeal stimulations are impaired during respiratory syncitial virus infection (Lindgren C et al., 1996).

Arousal thresholds decline across the night, as a function of accumulated sleep time (Rosenthal L et al., 1996; Franco P et al., 2001). Infants respond poorly when exposed to auditory challenges during the first sleep cycle of the night, and arousal thresholds decrease across the night, independently of sleep stage (Busby KA et al., 1994; Rosenthal L et al., 1996). In infants born at term, arousal thresholds to pulsatile air jet applied to the nostrils increased progressively with time of quiet sleep (Read PA et al., 1998).

Spontaneous awakenings occur more frequently in infants sleeping in active/REM sleep than in quiet/NREM sleep. Lower arousal thresholds are seen during active sleep in response to diverse stimuli, such as pharyngeal (Jeffery HE et al., 1999), auditory (Rosenthal L et al., 1996; Franco P et al., 2001), vibrotactile (Newman NM et al., 1989), pulsatile air jet to the nostrils (Read PA et al., 1998), visual (Kahn A et al., 2000), airway occlusion (Thach BT et al., 1989; Berry RB, Gleeson K, 1997), or head-up tilt challenges (Galland BC et al., 1998). Responses to hypoxia during nighttime sleep or during active sleep should be investigated in infants. In children, arousal occurred less during deep NREM sleep than in stage 2 NREM sleep and REM sleep for hypoxic and hypercapnic challenges (Marcus CL et al., 1998b).

The rapid repetition of the stimuli can favor the development of habituation and modify the sleeper's arousal threshold responses. It was shown that rapid habituation of responses occur with repeated tactile stimulation to the foot, performed during daytime nap studies (McNamara F et al., 1996).

Some studies reported significantly higher arousal thresholds in newborns and infants pre-natally *exposed to nicotine and cigarette smoke*. The finding was reported for newborns and infants of mothers who smoked during gestation. When exposed to auditory (Franco P et al., 1999), hypoxic (Lewis KW, Bosque EM, 1995) arousal challenges, these newborns and infants had significantly higher arousal thresholds than subjects born to non-smoking mothers. The newborn's arousability is also depressed following *maternal exposure to illegal drugs*. Infants of substance-abusing mothers required a significantly longer exposure to hypercapnia before arousal than age-matched control infants (Davidson Ward SL et al., 1992b). A preliminary study reported that disturbances in spontaneous arousals occurred in neonates of mothers who consumed alcohol during the first trimester (Scher MS et al., 1988).

Arousal thresholds depend also of the *age of the infant*. Like spontaneous awakenings (Thomas DA et al., 1996), arousal thresholds to auditory stimuli (Kahn A et al., 1986c) show an ontogenetic evolution and decrease from birth during the first 3 months of life. As normal infants mature, their ability to arouse in response to hypoxia diminishes by 2-3 months of life. Indeed, only 12.5% of infants older than 9 weeks of age and 70% of infants younger than 9 weeks of age arouse (Davidson Ward SL et al., 1992a). However, by the same age, an increase in vibrotactile stimuli arousal thresholds was reported (Newman NM et al., 1989).

The duration of the gestation can modify the infant's arousability. At 2-3 months of age, the arousal thresholds are greater in pre-term infants challenged by nasal air-jet stimulations during NREM sleep, than in infants born at term (Horne RSC et al., 2000). Moreover, pre-terms with a history of apnea and bradycardia of prematurity showed decreased responses in both quiet sleep and active sleep. Arousal in infants are accompanied by specific behavioral stereotypes. Four major behavioral features associated with arousals have been described: sighs startles, thrashing, and full awakening with eyes opening and sometimes cry. These behavioral reactions usually occur in the infants during arousals in the following order: first, deep sighs and startles happen, then thrashing with wide movements of the extremities may be seen and, finally, the infant wakes up (Lijowska AS et al., 1997). At the same time, listed behavioral reactions may be incomplete.

In adult subjects, *sleep deprivation* significantly increases the pressure to sleep. The arousal threshold is essentially increased during the second part of the night, in both REM and in NREM sleep (Rosenthal L et al., 1996). Following short-term sleep deprivation in infants, decrease in spontaneous movements and increase in arousal thresholds were found after auditory challenge in active sleep, but no detectable alteration was found after light and auditory stimuli in quiet sleep (Thomas DA et al., 1996). It is thus conceivable that sleep deprivation from any cause, including upper respiratory infection, could result in diminished arousal responsiveness. Sleep deprivation can result from nursing conditions (i.e., handling, feeding) as well as from sleep fragmentation due to respiratory infections or fever (Franco P et al., 1999). On the day of the discharge from hospital after either a respiratory or urinary infection, infants showed decreased arousability after nasal air jet stimuli in quiet sleep when compared to when they were completely well 10-15 days after discharge (Horne RSC et al., 2002). Some medications as antitussive drugs can also diminish the frequency of spontaneous movements and decrease arousability (Kahn A et al., 1985a).

Body position during sleep significantly modifies the arousal responses in infants. Newborn and 3-month old infants showed less arousals (Kahn A et al., 1993a) and higher auditory arousal thresholds (Franco P et al., 1998a; Galland BC et al., 1998; Groswasser J et al., 2001; Horne RSC et al., 2001) than when lying supine. The sleep-promoting effect of the prone sleep position is independent of the infant's usual sleep position, and remains unexplained.

Newborns *exposure to cold* showed a decrease in sleep continuity and an increased frequency of active sleep and body movements (Bach V et al., 1994). Compared to what was observed during thermo-neutrality, newborns sleeping in a warm environment had higher arousal thresholds to auditory stimuli, mainly by the end of the night (Franco P et al., 1999). Infants sleeping with their face covered by a bed sheet had higher auditory thresholds, greater rectal and pericephalic temperatures (Franco P et al., 2002b). These increase arousal thresholds have been related to an elevation in temperatures within the infant's microenvironment.

Arousals thresholds occurred at significantly lower auditory levels in infants *sleeping with a pacifier* than in infants who do not use a pacifier (Franco P et al., 2000a), and in subjects *breast-fed* than in infants bottle-fed (Elias ME et al., 1986; Franco P et al., 2000a). *Bed-sharing* also favors repeated arousals (Mosko S et al., 1995).

Tables 33 and 34 summarize the data on the major modifying factors influencing arousal in the infants and young children.

Table 33. Factors that decrease the infant's arousability from sleep, seen as changes in spontaneous or induced arousals

Factors	Effect	References
Prenatal:		
Drugs	Delayed arousal response to hypoxia	(Davidson Ward SL et al., 1992b)
Smoking	Reduced arousal response to hypoxia	(Lewis KW, Bosque EM, 1995)
	Higher auditory threshold	(Franco P et al., 1996)
	Decreased arousability after nasal air-jet	(Franco P et al., 1996)
Alcohol	Decreased spontaneous arousals	(Scher MS et al., 1988)
Postnatal:		
Increase in age	Decrease in spontaneous arousals	(Kahn A et al., 2000)
	Higher auditory arousal threshold	(Kahn A et al., 1986c)
	Higher threshold to trigeminal nerve stimulation	(Horne RSC et al., 2000)
	Higher threshold to hypoxic arousal	(Davidson Ward SL et al., 1992a)
	Higher threshold to vibrotactile arousal	(Newman NM et al., 1989)
Sleep deficiency (deprivation)	Higher auditory arousal threshold	(Rosenthal L et al., 1996)
Sedatives	Decrease in spontaneous arousals	(Kahn A et al., 1985a)
Respiratory Infections	Decrease arousability after nasal air jet	(Horne RSC et al., 2002)
Experimental:		
Time of the night	Changes in auditory threshold	(Franco P et al., 2001)
Prone sleep position	Decrease in spontaneous arousals	(Kahn A et al., 1993a)
	Decreased arousability after auditory test	(Franco P et al., 1998a)
	Decreased reaction to head-up tilt-test	(Galland BC et al., 1998)
	Decreased reaction to nose air-jet	(Horne RSC et al., 2001)
	Decreased reaction to obstructive apnea	(Groswasser J et al., 2001)
Higher room temperature	Higher auditory threshold	(Franco P et al., 1999)

Arousal in response to respiratory disorders has a protective effect. At the same time, as a consequence, multiple arousals may lead to fragmented and inefficient sleep in infants and children with sleep disordered breathing. In turn, this may lead to functional disorders in the day time, in particular to drowsiness. Even minor influences leading only to transitory increase in arterial blood pressure, so-called "vegetative arousals", may result to hypersomnia during the day time (Martin SE et al., 1997). This is equally true for the adults and children, although in the latter day time hypersomnia found against a background of OA in sleep is not that pronounced (Davidson Ward SL, Marcus CL, 1996). Partly, this finding may reflect the lack of cortical arousals resulted from hypoxemia. OA occurring during quiet sleep in the infants and young children end up with arousals in only 39 % observations; in the active/REM-sleep this is noticeable in 38% cases. OA more often than CA lead to arousals in the infants, and the chance of an infant to terminate apneic spell during sleep by efficient arousal increases with infant's growth and development (McNamara F et al., 1996). Infants suffering OA during sleep have less proportion of the active/REM sleep compared with the healthy babies. However, the total number of arousals observed against a background of OA does not markedly differ from that found in the practically healthy children. Hence, sleep architecture in the infants and young children with OA is not altered to such an extent, as it may be the case in the adult patients.

Table 34. Factors that increase infant's arousability from sleep, seen as changes in spontaneous or induced arousals

Factors	Effect	References
Infant factors:		
Use of pacifier	Increased spontaneous arousals	(Elias ME et al., 1986)
	Lower auditory arousal threshold	(Franco P et al., 2000a)
Breast feeding	Increased spontaneous arusals	(Elias ME et al., 1986)
	Lower auditory arousal threshold	(Franco P et al., 2000a)
Milk allergy	Increased spontaneous arousals	(Kahn A et al., 1985b)
Prolonged apneas	Increased spontaneous arousals	(Rees K et al., 1995)
Gastroesophageal reflux	Increased spontaneous arousals	(Kahn A et al., 1991b)
Periodic leg movements	Increased spontaneous arousals	(Winkelman JW, 1999)
Pain	Increased spontaneous arousals	(Oberlander TF et al., 1999)
Experimental:		
Infant-parent bed sharing	Increased spontaneous arousals	(Mosko S et al., 1995)
Tracheobronchial stimuli	Increased spontaneous arousals	(Jeffery HE et al., 1999)
Increased blood pressure	Increased spontaneous arousals	(Trelease RB et al., 1985)

Part 2. Major Forms of Sleep Disorders in Infants and Young Children and Approaches to Their Correction

Various forms of sleep disorders may be recognized in infants and young children. According to international classification, following major types of sleep disorders should be distinguished (Committee DCS, 1990):

1. Disorders of initiating and maintaining sleep (DIMS), or *insomnia*s
2. Disorders of excessive somnolence (DOES).
3. Disorders of circadian rhythm of sleep-wake cycle.
4. Disorders of sleep connected with the changes in its structure, separate stages, partial awakenings occurring after the sleep onset, known as *parasomnias*

Insomnia and excessive somnolence are often combined by the common term – *dissomnias*. Dissomnias in turn may be divided into *external*, developing under influence of exogenous factors (disorders of sleep hygiene, influence of medications), and *internal,* caused by biological factors.

There are data that 88 % of parents' consultations related to sleep disorders in children under 2 years old are linked to insomnia (Navelet Y, 1996). By the end of the first year of a life, night wakes become increasingly common, sometimes accompanied by long vigilance: they are encountered in 10 - 44 % infants (Adair RH, Bauchner H, 1993). The most common parental concern is the infant's bad sleep, long periods of wakefulness at night, as well as morning awakenings in the early hours (Anders TF et al., 1992). By the end of the second year of life, such parental complaint as infant's active resistance to being put to bed at the bedtime is not uncommon. This behavior is observed in 8 % children aged 15 months to 4 years. Children of the specified age group exhibit the signs of insomnia in 42 % cases (Kataria S et al., 1987).

Generally, the cause of internal insomnias is some kind of imbalance between the activity of the brain structures normally responsible for the maintenance of wake and sleep states, and certain forms of such insomnias may be distinguished (Glaze DG et al., 2002). One may speak about transitory and situational insomnias, if their duration does not exceed three weeks. Often they may be caused by known stressful situation. On the contrary, persisting psycho-physiological insomnia assumes involvement of the organic causes leading to sleep disorders and some situational factors that may aggravate this disorder. Insomnias in young children often associated with food allergy (Kahn A et al., 1989; Stores G et al., 1998), with gastro-esophageal reflux (Kahn A et al., 1991b; Ghaem M et al., 1998), with so-called colic (Weissbluth M, 1995b). Insomnia may become the sign of serious neurological abnormality and progressive mental diseases. Particularly, it is important to note frequent associations between insomnias and respiratory disorders during sleep: such association is characteristic both for central and obstructive apneas (Guilleminault C, Dement WC, 1978; Kelmanson IA, 1997). In some children suffering insomnia sleep disordered breathing does not manifest as apnea, but the breathing becomes shallow leading to reduced level of blood oxygen saturation, known as *hypopnea.* This type of breathing disorder most commonly occurs during active/REM-sleep.

Much interest has been raised in sleep disorders in the infants born pre-term and (or) low birth weight. This issue has not been addressed extensively so far, although there are the reasons to believe that this category of infants should have certain peculiarities of sleep. The findings were that these babies more frequently needed the parents to be present in the infant's room and to be put in parents bed to fall asleep (Kelmanson IA, Adulas EI, 2002).

The insomnia in the infants may occur against a background of *nocturnal myoclonus and restless legs syndrome* (Coleman RM, 1982). Although the diagnosis of restless legs syndrome is usually reserved for the adults, a closer look at these patients enables to certify that in 38-45 % cases initial signs of disorder are noticeable as early as childhood (Walters AS et al., 1996), and restless legs syndrome is diagnosed in 17 % children aged of 2-14 years (Hornyak M et al., 2002). Restless legs syndrome and (or) periodic limb movement often accompany other somatic disorders. In particular, this can be found in young children suffering OSA (Ali N et al., 1996; Chervin RD, Archbold KH, 2001; Scholle S et al., 2001), *juvenile fibromyalgia* (Tayag-Kier CE et al., 2001) and *Tourette disorder* (Lipinski JF et al., 1997). It is important to pay attention to the proved connection of the restless legs syndrome and (or) periodic limb movement with the signs of day time hyperactive behavior and attention deficiency in children (Picchietti DL et al., 1998; Picchietti DL et al., 1999; Hornyak M et al., 2002). Restless legs syndrome has much in common with so-called *growing pains*. In the latter case, pains in the extremities are also observed during night and accompanied by sleep disorders (Oster J, Nielsen A, 1972). Opinions on similarities of the specified conditions are inconsistent: both skeptical (Ekbom KA, 1975), and affirmative (Brenning R, 1960) arguments were offered.

It is important to stress that insomnias are often associated with certain forms of day-time behavioral changes, and that the infants presented with day-time behavioral disorders need a closer look for potential sleep abnormalities. Most studies in infants and young children addressing the influence of lack of sleep on their day-time behavior and mood are related to the effects of apneas and other forms of sleep disordered breathing (Guilleminault C et al., 1982b; Ali NJ et al., 1993; Lewin DS et al., 1996; Hansen DE, Vandenberg B, 1997; Kelmanson IA, 2000a). Sleep disorders found in the infants and children as a results of apneas are accompanied by changes of mood (negativism), sleepiness during day-time, attention disorder, and anxiety. Sleep disorders may be accompanied by reduced periods of selective attention, reduced activity and memory deficiency (Lewin DS et al., 1996; Hansen DE, Vandenberg B, 1997; Mindell JA et al., 1999).

It is important to note that sleep disorders revealed in the infants on the first year of life may persist with advancing age. More than 40 % infants who have sleep disorders at the age 8 months also have sleep disorders at the age of 3 years (Zuckerman B et al., 1987). Reportedly, 84% infants who suffered sleep disorders at the age of 15 to 48 months had such findings by the age of 3 years, while only 3 % infants without the signs of sleep problems at initial inspection were found to have these problems when 3-years-old (Kataria S et al., 1987). Three-month-old infants with sleep disorders are characterized by specific day-time behavioral features; they were reported to be more fussy and whimsy.

It is possible to distinguish certain forms *of external insomnias* (Glaze DG et al., 2002).

1. *Insomnia due to disadaptation.* Extrinsic psychosocial stressors may have the potential to cause either acute or chronic sleep disruption. Emotional factors are critical, and as soon as the crisis is over, sleep disorder is likely to disappear. Many chronic medical conditions may be associated with difficulty settling, particularly neurological conditions such as cerebral palsy, autism, and other developmental disabilities. Many commonly used medications, including stimulants and some anticonvulsants, are known to cause insomnia in at least some children. Acute febrile illnesses such as otitis media will often cause short-term disruption of sleep onset or continuity.
2. *Insomnia due to regime violations.* This form usually manifests as a reluctance of a child to go to bed. This form of sleep disorder usually becomes obvious in 1 to 10-year-old children and normally disappears with improvement of child-rearing practices and optimization of sleep organization (Mindell JA, Durand VM, 1993). Stereotypes of infant's sleep behavior markedly depend on numerous external influences, and infant-parent interaction is one of the most potent. Extremely liberal or, on the contrary, unduly strict regime that regulates the time infant/child is put to bed, parental behavior and reaction on infant's night wakes, parental mood and mental health may be critical. As well, feeding patterns and duration, practice to put the baby into parental bed may markedly affect infant's behavior during sleep. According to own findings, certain features of infant's microenvironment may modify his/her behavior during night sleep. In particular, the babies facing more developmental stimulation and from more organized environment less often fell asleep in parents bed. Infants from more organized environment were more often put to sleep at the same time at night, more often were ready to go to sleep at bedtime and less often struggled at bed; it was more common with them to have right amount of sleep and to have about a same amount of sleep each day; less often they moved a lot during sleep and woke up in sleep (Kelmanson IA, Adulas EI, 2004). That is why the infants with parentally reported sleep problems should be carefully considered for possible flaws in the quality of environment. Consistent, well-structured routines leading to bedtime aid children in transitioning from their daytime level of activity and alertness to a quiet, relaxed state more conducive to sleep onset. Conversely, irregular schedule or highly stimulating pre-bedtime

activities often tend to disrupt sleep onset. In a similar fashion, a quiet sleep environment that is dark or only dimly lit will for most children tend to promote faster and more comfortable sleep onset.

3. *Sleep-onset association disorder.* It is a condition affecting primarily infants and younger children in which sleep onset becomes dependent on an environment or circumstance that cannot be independently recreated by the child without assistance. Potentially problematic associations for infants include getting dummies, being rocked, nursed, or bottle-fed at sleep onset. For toddlers, requiring parental presence at sleep onset is a common association. Although providing the specific circumstance desired by the child is usually perceived by parents to improve sleep onset at bedtime, some children become so reliant on the associated circumstance that they are unable to fall asleep in its absence. As a result, affected children are unable to settle themselves following the brief physiological awakenings that are part of normal sleep. The child, reliant upon the parent to provide the circumstance necessary for transition back to sleep, will cry or otherwise signal the parent, leading to frequent problematic awakenings during the night. This form of extrinsic insomnia is encountered in 46% of the visits to the pediatric clinic during the second 6 months of life, usually manifests in the infants aged 6 months to 3 years, and in most cases disappears with infant's growth (Mindell JA, Durand VM, 1993). Beginning after approximately 9 months of age, many infants experience distress when separated from their parents or caregivers. Although separation anxiety is considered a normal element of early childhood development, it can also precipitate significant distress at bedtime or during nighttime waking. This is typically self-limited and subsides with advancing age. Use of transitional objects such as blankets or stuffed animals may assist children in mastering separation responses that disrupt sleep.
4. *Persistent nighttime feedings.* In most healthy infants over 6 months of age, nighttime feeding is more likely to represent a learned behavior than a biological necessity. This is typically a self-limited behavior that subsides spontaneously as an infant matures. In situations where persistent nocturnal feedings significantly disrupt the quality and continuity of nighttime sleep, structured schedules for gradually reducing the quantity and frequency of nighttime feedings are usually highly effective.

The most common form of excessive somnolence is *narcolepsy.* Narcolepsy is a syndrome of unknown etiology characterized by excessive daytime sleepiness that is often profound. Narcolepsy is usually diagnosed in the adults, less often in teenagers, and is rather seldom observed in younger children. At the same time, there are strong reasons to believe that the rarity of the diagnosis of narcolepsy in the young age in mainly due to unsatisfactory recognition of this disorder. According to some reports, in 13 % adult patients with diagnosed narcolepsy, retrospective analysis could distinguish clinical symptoms highly suspicious to this condition in their childhood (Navelet Y et al., 1976). In approximately 20 % children who suffered narcolepsy diagnosed at the age 2 to 12 years, "hyperkinetic syndrome" has been reported when they have been younger, and the symptoms described were cataplexy and increased day time sleepiness (Guilleminault C, Pelayo R, 1998). It is not uncommon with younger children suffering narcolepsy to be misdiagnosed as epilepsy and getting improper treatment. The first signs of narcolepsy in children younger than 5 years old may become behavioral problems, particularly increased irritability, aggression, social isolation. Child inactivity and laziness may become common parental complaints in cases on preschool children with narcolepsy. Narcolepsy in young children may be associated with other forms of sleep-related disorders. In particular, up to 50 % patients with narcolepsy may suffer obstructive sleep apneas (Aldrich MS, 1992). Episodes of obstructive sleep apneas lead to disruptive sleep and, as a results, increased day-time sleepiness. As high as 40 % patients with narcolepsy have the signs of restless legs syndrome (Boivin DB et al., 1993).

Narcolepsy usually occurs in association with cataplexy and other symptoms and signs, which commonly include hypnagogic hallucinations or hypnopompic hallucinations, sleep paralysis, automatic behavior, and disrupted nocturnal sleep (Committee DCS, 1990). Symptoms most often begin during adolescence or young adulthood; however, narcolepsy may also occur earlier. The excessive daytime sleepiness of narcolepsy presents with an increased propensity to fall asleep, nod or doze easily in relaxed or sedentary situations, or a need to exert extra effort to avoid sleeping in these situations. Additionally, irresistible or overwhelming urges to sleep commonly occur from time to time during wakeful periods in the untreated patient with narcolepsy. These so-called "sleep attacks" represent the episodes of profound sleepiness experienced by those with marked sleep deprivation or other severe sleep disorders. In addition to frank sleepiness, the excessive daytime sleepiness of narcolepsy, as in other sleep disorders, can cause related symptoms including poor memory, reduced concentration or attention, and irritability. Cataplexy is the partial or complete loss of bilateral muscle tone in response to strong emotion. Reduced muscle tone may be minimal, occur in a few muscle groups, and cause minimal symptoms such as bilateral ptosis, head drooping, slurred speech, or dropping things from the hand. Or, it may be so severe that total body paralysis occurs, resulting in complete collapse. Cataplectic events usually last from a few seconds to 2 or 3 minutes but occasionally continue longer. The patient is usually alert and oriented during the event despite the inability to respond. Positive emotions such as laughter more commonly trigger cataplexy than negative emotions; however, any strong emotion is a potential trigger. Startling stimuli, stress, physical fatigue, or sleepiness may also be important triggers or factors that exacerbate cataplexy. According to epidemiological studies, cataplexy is found in 60 to 100% of patients with narcolepsy. A large range in percentage affected with cataplexy is reported because definitions of narcolepsy vary among studies. The onset of cataplexy is most frequently simultaneous with or within a few months of the onset of excessive daytime sleepiness, but in some cases, cataplexy may not develop until many years after the initial onset of excessive daytime sleepiness. Another phenomena related to narcolepsy are hypnagogic or hypnopompic hallucinations. They may be visual, tactile, auditory, or multisensory events, usually brief but occasionally continuing for a few minutes, that occur at the transition from wakefulness to sleep (hypnagogic) or from sleep to wakefulness (hypnopompic). Hallucinations may contain elements of dream sleep and consciousness combined and are often bizarre or disturbing to patients. Sleep paralysis is the inability to move, lasting from a few seconds to minutes, during the transition from sleep to wakefulness or from wakefulness to sleep. Episodes of sleep paralysis may alarm children, particularly those who experience the sensation of being unable to breathe. Although accessory respiratory muscles may not be active during these episodes, diaphragmatic activity continues and air exchange remains adequate. Other commonly reported symptoms include automatic behavior ("absent-minded" behavior or speech that is often nonsensical and that the patient does not remember) and fragmented nocturnal sleep (frequent awakenings during the night).

Hypnogogic hallucinations, sleep paralysis, and automatic behavior are not specific to narcolepsy and occur in other sleep disorders, as well as in healthy individuals. These symptoms are, however, far more common and occur with much greater frequency in narcolepsy.

Nocturnal PSG is an important part of the evaluation process primarily to exclude other conditions that occur in narcolepsy at a higher than normal rate (obstructive sleep apnea, periodic limb movement syndrome, and rapid eye movement (REM) sleep behavior disorder) and could add to the sleepiness or nocturnal sleep disruption the patient may be experiencing. Daytime nap studies, in the form of the multiple sleep latency test (MSLT), usually demonstrate substantially reduced sleep latency and sleep onset REM periods in patients with narcolepsy. Sleep cycle begins with the episode of REM-sleep (Aldrich MS, 1992; 1998). Meanwhile, sleep onset REM periods are not specific for narcolepsy, but the occurrence of two or more of these events during the MSLT, in the setting of objective marked sleepiness and without another explanation for their occurrence (e.g., sleep deprivation, REM-suppressant medication rebound, altered sleep schedule, obstructive sleep apnea, or delayed sleep-phase syndrome) is suggestive of narcolepsy. However, this may have diagnostic value only for the children of older age and adults, since in the healthy infants and young children sleep also begins with REM periods.

The similarity between cataplexy and REM sleep atonia, the presence of frequent episodes of hypnagogic hallucinations and of sleep paralysis, and the propensity for adult narcoleptics to go directly from wakefulness into REM sleep suggest that narcolepsy is primarily a « disease of REM sleep ». According to this hypothesis, a cataplectic attack represents an intrusion of REM sleep atonia during wakefulness, and the hypnagogic hallucinations appear as dreamlike imagery taking place in the waking state, especially at sleep onset in patients who frequently have sleep onset REM.

Cataplexy is associated with an inhibition of the monosynaptic and polysynaptic deep tendon reflexes, which is normally typical only for REM sleep (Guilleminault C et al., 1974). Although cataplexy and REM sleep atonia have great similarity and possibly share a common executive system, it is not necessary for the regulatory mechanism of both states to be identical. The mechanism of emotional triggering of cataplexy remains undetermined.

Much research into the HLA association in narcolepsy has been conducted. The most specific marker of narcolepsy in several different ethnic groups studied to date is DQB1*0602 (Mignot E, 1998). The strong association between HLA type and narcolepsy with cataplexy raises the possibility that narcolepsy is an autoimmune disease.

Attention was paid to the role of *hypocretin/orexin* system in sleep regulation and its potential involvement into the origins of narcolepsy. These peptides were identified in the late 1990-ies and were called "hypocretin" because of the primary hypothalamic localization and similarities with the hormone secretin, and were called "orexin" by other researchers after observing that central administration of these peptides increased appetite in experimental animals. Hypocretin-1 and -2 are produced exclusively by a well-defined group of neurons localized in the lateral hypothalamus. The neurons project to the olfactory bulb, cerebral cortex, thalamus, hypothalamus, and brain stem, particularly the locus coeruleus, raphé nucleus, and to cholinergic nuclei and cholinoceptive sites (such as pontine reticular formation), thought to be important for sleep regulation (Peyron C et al., 1998). Hypocretin system is a major excitatory system that affects the activity of monoaminergic (dopamine, norepinephrine, serotonin, and histamine) and cholinergic systems with major effects on vigilance states. It is thus likely that a deficiency in hypocretin neurotransmission induces an imbalance between these classical neurotransmitter systems, with primary effects on sleep-state organization and vigilance. Hypocretin levels increase during the active periods and are highest at the end of the active period, and the levels decline with the onset of sleep. How hypocretin neuronal activity varies across different sleep stages is still unknown. Regardless of the firing pattern of the hypocretin neurons, the findings suggest that basic hypocretin neurotransmission fluctuates across the 24-hour period and slowly builds up during the active period. Due to the lack of increase in hypocretin tone, narcoleptic subjects may not be able to stay awake for a prolonged period and do not respond to various alerting stimuli. Conversely, the reduction of the hypocretin tone at sleep onset may contribute to the profound deep sleep that normally inhibits REM sleep at sleep onset, and the lack of this system in narcolepsy may allow the occurrence of REM sleep at sleep onset.

Certain forms of *parasomnias* have been described in children:

1. *Confusional arousals***.** Confusional arousals occur in the first 3 hours after sleep onset when the child experiences an abrupt transition from the deepest phases of NREM sleep, stages 3 and 4, to a lighter stage of sleep. This transition can be associated with minor episodes of mumbling or grimaces, disorientation, or inappropriate behaviors (Mindell JA et al., 1999). For the most part, the child remains asleep throughout the episode and has no recall of the occurrence the next day. Most episodes are brief, averaging 2 to 10 minutes in duration, and end when the child re-enters deep sleep (Rosen G et al., 1995). Predisposing factors for confusional arousals include obstructive sleep apnea, overtiredness, sleep deprivation, and irregular wake/sleep schedules, all of which alter patterns of stage 3 and 4 NREM sleep (Rosen G et al., 1995). Prevalence rates are most likely underestimated because children have no recollection of the event and parents often do not observe these nighttime occurrences. Parents should be reassured that these episodes are harmless and that they should not awaken their child. Confusional arousals diminish as the child matures.
2. *Sleepwalking***.** Sleepwalking is common, occurring in 1% to 15% of children, and usually begins when the child is between the ages of 4 and 8 years. In approximately 6 % cases, these disorders persist across the time (Soldatos CR, Lugaresi E, 1987; Anders TF et al., 1992; Mindell JA, 1993; Mindell JA, Durand VM, 1993), and sometimes can persist into late adolescence and adulthood (Ferber R, 1996). A sleepwalking episode consists of minor actions such as sitting up in bed, walking around the room or house, or other activities. Because the sleepwalker

may try to unlock doors or windows and go outside, precautions must be taken to ensure the safety of the child, such as locking doors and windows and placing barriers in front of stairwells. The sleepwalker does not meaningfully interact with people and is often easily agitated. Arousal is difficult and, if successful, the child will appear confused. Recollection of the event is rare (Adair RH, Bauchner H, 1993). Management includes informing the parent that the child is truly asleep and should be gently redirected back to bed without awakening. Parents should intervene, however, if the child is in danger. Parents may place a bell on the child's bedroom door to alert them whenever he leaves the bedroom, and locks placed out of reach on outside doorways may be necessary. Extremely agitated sleepwalkers may benefit from sleeping on the first floor of the house.

3. *Night terrors.* Night terrors are reported in 1% to 6% of the pediatric population, are generally seen after 18 months of age, more often in boys, and usually disappear by age 6 years (Soldatos CR, Lugaresi E, 1987; Mindell JA, 1993; Mindell JA, Durand VM, 1993; Anders TF, Eiben LA, 1997; Mindell JA et al., 1999). Night terrors are typified by an extremely agitated child suddenly aroused from stage 4 sleep and are often accompanied by screaming, increased heart rate, and dilated pupils. Children appear confused and are usually inconsolable (Dahl RE, 1998). Children experiencing a night terror appear awake, but are not; they will not coherently interact with people or their surroundings. During an episode, children are extremely difficult to arouse and will not remember the event if aroused or upon awakening in the morning (Adair RH, Bauchner H, 1993). In fact, it is advisable to gently direct children away from danger (stairs, windows, doors to the outside) and back to their own beds. Children will then promptly fall back to sleep. Children with more violent, prolonged, or unusual episodes should be evaluated at a pediatric sleep disorders center. Although night terrors are often very disturbing to parents, they can be reassured that night terrors are a developmental phenomenon and are not a sign of underlying psychopathology.
4. *Nightmares.* The incidence of nightmares ranges from 25% to 50% of children aged 3 to 6 years, and both sexes are equally affected (Adair RH, Bauchner H, 1993). Nightmares are frequently confused with sleep terrors, but there are fundamental differences between the two (Table 35). Nightmares occur during REM sleep and therefore are seen during the last half of the night. Children become fully awake after a nightmare, can remember graphic details, and frequently remember and talk about their nightmare the next morning (Mindell JA et al., 1999). Occasional nightmares are not worrisome, but recurrent nightmares or those with disturbing content may indicate excessive daytime stress (Adair RH, Bauchner H, 1993). Clinicians and parents should gently question children regarding their daytime experiences such as television shows or encounters with other children to try to determine possible frightening content or situations. Once the basis for the nightmares is discerned, measures should be taken to eliminate or reduce the child's exposure to the causative factor.
5. *Rhythmic movement disorder.* Rhythmic movement disorder involves the rhythmic movement of large muscle groups, such as the head and neck, and occurs when the

child is drowsy, attempting to fall asleep, and can continue into stage 1 sleep. The first incidence usually occurs before the second birthday; headbanging and body-rocking are the most common types. PSG monitoring during an episode can rule out a medical cause such as seizure activity (AASM, 2000). The disorder is usually transient and self-limited and seldom requires treatment (Hoban TFR, 2003).

6. *Bruxism* (teeth grinding). It becomes apparent in approximately 50 % younger children. The mechanisms, potential consequences and approaches to treatment are studied insufficiently so far.
7. *Enuresis* (bed wetting). This form of parasomnia may be definitely conformed after the child is 5 years or older. Approximately 1 of 10 children aged 6 years and 1 of 20 children aged 10 years suffer enuresis; boys more often than girls (Mindell JA, 1993). Obviously, it is difficult to discuss that form of parasomnia with the reference to infants and younger children, although there are good reasons to believe that this disorder featuring later in life has its roots in the early age.

There is increasing evidence that sleep problems can have a negative effect on infants and children's cognitive, behavioral, and emotional functioning. Most young children do not have typical symptoms of sleep loss such as yawning or obvious sleepiness that are observed in adolescents and adults. Rather, children often manifest symptoms that are paradoxical to sleepiness such as irritability, hyperactivity, short attention span, and low tolerance for frustration (Dahl RE, 1996). These indicators are often overlooked because children may have a long history of these symptoms or they may be mild in nature (Ferber R, Kryger M, 1995). Several studies have found a link between typical symptoms of inadequate sleep in children and symptoms characteristic of attention deficit/hyperactivity disorder (ADHD) (Dahl RE, 1996). Children diagnosed with ADHD have been shown to have a higher incidence of sleep problems and shorter average sleep duration than children without ADHD. Sleep loss has been implicated in childhood injuries.

Table 35. Major differences between night terrors and nightmares (Ferber R, 1985)

Signs and symptoms	Night terrors	Nightmares
Average age of onset	>18 mo	36-72 mo
Timing	First third of night	Last half of night
Sleep stage involved	NREM	REM
Agitated appearance of child	Yes	Sometimes
Child awakens or consolable	No	Yes
Returns to sleep easily	Yes	No
Recall of event	No	Yes

Evaluation of sleep problems involves a thorough history of the child's 24-hour routine focusing on bedtime habits, nighttime behaviors, naps, and daytime behavior. All behaviors and activities need to be evaluated for appropriateness for the child's age and time of day, recognizing the opportunity to make beneficial changes to restore optimal sleep for both the child and family. Nighttime behaviors after sleep onset such as confusional arousals, night terrors, nightmares, number and duration of night wakings, and symptoms of obstructive sleep apnea (ie, snoring, gasping, breathing pauses, and restless sleep) also should be assessed. Daytime behaviors should be reviewed with the parent. Also, stress within the family can negatively affect a child's sleep (Mindell JA et al., 1999). Pediatric clinicians rarely ask questions pertaining to sleep during well-child visits. Additionally, many parents may fail to bring up issues surrounding sleep problems to the health care provider, so direct questioning is the key to initiating a discussion (Owens JA et al., 2000). A sleep diary can be helpful in delineating the exact sleep pattern. Sleep diaries collect information regarding bedtime, frequency and duration of nighttime awakenings, nap frequency and duration, time of morning waking, and total sleep time. A thorough physical examination should be performed on any infants and child with a sleep complaint to assess for medical causes. Most sleep problems are alterations of normal sleep behavior created by poor sleep hygiene, developmental sleep phenomena, or stressful daytime activities and schedules (Adair RH, Bauchner H, 1993).

Information on infant's and child's behavior during night should be completed by the features of his/her day-time behavior. Of particular value may be use of the standardized questionnaires aimed at estimation of the qualitative behavioral features (infant's *temperament*). Although during wakefulness, temperament characteristics become more sharply focused and measurable, it has been argued that there is no true independence between these two state-specific behaviors (Weissbluth M, 1989). This may seem particularly true for the very young infants in whom neurological states called "sleep" and "wakefulness" are relatively indifferent (Hoppenbrouwers T et al., 1988). Specific infant behavioral styles were implicated to certain sleep disturbances (Table 36).

Table 36. The features of temperament associated with sleep troubles in infants (Weissbluth M, 1989)

Temperament feature	Parental estimates	Sleep impairments
Irregularity	Different time the infant falls asleep	Violations of sleep regime
High intensity and negative mood	Continuous and unmotivated infant cry during day time	Night awakenings
Slow adaptability	Inability to establish strict schedule of infant's sleep during day time	Insufficient day time sleep
Low threshold (high sensitivity to external stimuli)	Increased propensity to wake up at night	Fragmented sleep
High persistence	Increased time the infant needs to be put to bed and fall asleep	Increased sleep latency

It was also shown that the infants with more negative mood more often required rocking or rhythmic movements to fall asleep, resisted sleeping alone, slept too little, were restless and moved a lot during sleep, woke up during night screaming, sweating and being inconsolable; less often could return to sleep after waking without parental interference. Infants with higher intensity less often fell asleep alone. More active infants less often slept about the same amount each day. Less distractible babies were less often ready to go to bed at bedtime, more frequently resisted going to bed at bedtime, were reluctant to sleep alone, slept too little and less often slept about the same amount each day. More rhythmic infants were more often ready to go to bed at bedtime and less often resisted going to bed at bedtime; less often they slept too little, were restless and moved a lot during sleep, woke up very early in the morning (Kelmanson IA, 2004). Such forms of infant's sleep impairments as frequent night awakenings requiring extra feedings, infant's awakenings in the early morning hours, his/her protest against day time sleep, difficulty of an infant to fall asleep in the new bed and in unfamiliar place are most clearly associated with infant's mood and rhythmicity (Keener MA et al., 1988). The infants with so-called difficult temperament were characterized by frequent night waking (Carey WB, 1974) and shorter sleep duration (Weissbluth M, 1981). Likewise, the mothers regarded colicky infants as more intense in their reactions and more negative in their mood (Lehtonen L et al., 1994). Infants who required care giving during the night were rated as significantly more difficult and arrhythmic (Keener MA et al., 1988).

It is difficult to determine whether disturbed sleep is caused initially by a specific temperament trait or by a parent's behavior (in response to a trait or independently) because interactive effects always evolve. Regardless of etiology, the end result of all sleep disturbance is sleep-loss stress associated with a predictable stimulating biochemical response. The sleep deficit causes chemically stimulating stress response in the infants which adversely affects mood, performance, and temperament (Weissbluth M, 1989). It was suggested that certain biological active substances may be involved in interrelationships among sleep behavior and temperament factors. Particular attention was paid to progesterone level, and it was assumed that elevated plasma progesterone levels may lead to CNS depression and blunted arousals for sleep (Weissbluth M, 1982b).

It is unclear whether sleep-temperament interactions may persist across the time. However, some authors provided affirmative arguments. Significant correlations between sleep measures in a neonatal period and later temperament dimensions have been reported (Novosad C et al., 1999), and the children presented with colicky behavior during sleep in the infancy displayed more negative emotions according to the temperament scale at four years of age (Canivet C et al., 2000). Insufficient sleep time was also a more common finding among older children with difficult temperament (Jiang F et al., 2002). Behavioral sleep disturbances in children were associated with intense and negative temperament characteristics (Owens-Stively J et al., 1997), and with "intermediate high to difficult" temperament profile (Atkinson E et al., 1995).

Treatment of sleeplessness in infants and younger children usually begins with effective bedtime management—a set of interventions focused on establishing a regular sleep schedule, an appropriate sleep environment, and consistent limit-setting for dealing with bedtime struggles (Ferber R, 1995a).

1. *Bedtime routine.* A regular and well-structured set of pre-bedtime activities usually helps settle children and promote sleep onset. The routine for many younger children often includes changing, bathing, stories, and being "tucked in."

Stimulating activities should be avoided due to their potential alerting effect. Keeping the routine quiet and regular helps young children achieve a quiet, relaxed state more conducive to sleep onset.

2. *Sleep schedule.* Maintaining a consistent bedtime and waking time 7 nights per week is usually helpful in several respects. Eliminating late sleeping and any daytime napping that is inappropriate for age may increase the chance that a child will be tired at the usual bedtime. In addition, regularity of sleep schedule may promote entrainment and stabilization of circadian rhythms as an added impetus for more rapid sleep onset at the regular bedtime (Ferber R, 1995b).
3. *Sleep environment.* For most infants and young children, a quiet environment is more conducive to sleep than a noisy one. Likewise, a dark or dimly lit environment promotes settling and sleep onset better than bright lighting, although use of a night-light is appropriate for children who are afraid of the dark. Finally, it is generally recommended that children be put to sleep without a parent remaining present. This provides the child with an opportunity to learn to fall asleep comfortably and independently and to minimize dependence upon parental presence or intervention as a sleep onset association.
4. *Limit setting.* Children who stall, cry, or leave the room at bedtime in an attempt to stay up later will sometimes repeat this behavior to the point where a parent or caregiver "gives in" and allows the child to stay up. For some children, this pattern of behavior may be repeated nightly to the point of causing consistent delay and disruption of sleep onset. Setting and enforcing appropriate limits on inappropriate bedtime behavior are crucial for the effective treatment of limit-setting sleep disorder (Ferber R, 1995a). Limits must be enforced consistently by all caregivers, usually for periods of days or weeks, for maladaptive bedtime behaviors to subside. Parents should be made aware that temperamental and agitated behavior often transiently worsen during the first days of treatment before gradual improvement becomes evident, making the initials days of treatment the most difficult.

In cases where bedtime management alone is not sufficient in the treatment of a younger child's sleeplessness, a variety of structured behavioral interventions are available (Ramchandani P et al., 2000).

1. *Extinction (systematic ignoring).* This technique involves placing a child in bed and ignoring agitation or inappropriate behavior until morning except for legitimate concerns regarding illness or safety of the child (Kuhn BR, Elliott AJ, 2003).
2. *Graduated extinction.* Using this technique, parents ignore bedtime tantrums or nighttime waking for a specified period of time before being allowed to enter the bedroom and briefly calm the child before exiting. The specified interval for ignoring the child's agitation progressively increases over time, usually over successive nights, and in some versions on successive checks within the same night (Ferber R, 1985; Kuhn BR, Elliott AJ, 2003).
3. *Scheduled awakenings.* For young children with nighttime waking occurring at predictable times during the night, this technique requires that the parent

preemptively awaken and then resettle the child 15 to 30 minutes before the usual time of spontaneous awakening. If effective in eliminating the spontaneous awakenings, the scheduled awakenings are gradually spaced out or delayed systematically. This technique is probably efficacious intervention for excessive night waking (Kuhn BR, Elliott AJ, 2003).

4. *Bedtime fading.* This technique initially delays a child's bedtime by ~30 minutes. If rapid sleep onset is not observed, a "response cost procedure" allows the child to be removed from bed and kept awake for 30 to 60 minutes, with the process being repeated as necessary until the child falls asleep rapidly. On subsequent nights, the initial bedtime is set 30 minutes before the time of rapid sleep onset was achieved on the night immediately prior, and the response cost procedure repeated if the child does not fall asleep quickly. The rules are applied on successive nights until the child's sleep onset is "faded" toward a predetermined bedtime goal (Kuhn BR, Elliott AJ, 2003).

Complementary therapy aimed at improvement of the quality of sleep in the infants may seem of particular interest. It was shown that LBW infants who received massage intervention less frequently required extra-feeding on waking-up at night, and were apparently more alert at the daytime (Adulas EI, Kelmanson IA, 2005). This was in accord with reports on the improved post massage sleep (Agarwal KN et al., 2000) and more alertness (Dieter JN et al., 2003) in the infants. It was found that tactile/kinaesthetic stimulation in pre term neonates was associated with more activity and alertness during sleep/wake behaviour observations (Field TM et al., 1986). Massage therapy reportedly enhanced the adjustment of circadian rhythms to the nocturnal period in the infants (Ferber SG et al., 2002).

Drug treatment of sleeplessness in infants and younger children has received limited scientific study. Children with insomnia were attempted to be given sedatives by 18 months of age (Ounsted M, Hendrick A, 1977). In a survey of pediatricians, psychiatrists, and primary care providers, sleep problems were the most common reason for prescribing medicine to infants up to 1 year of age, accounting for 35% of prescriptions in this age group (Trott GE et al., 1995). *Benzodiazepines* are used in pediatric patients for the treatment of night terrors (Mindell JA, Durand VM, 1993). They are also helpful in cases of confusional arousals, since reduce the number of arousals (Dahl RE, 1992; Ferber R, Kryger M, 1995). However, these drugs may only mask the symptoms of sleep disorder rather than really improve its quality; besides, they may lead to reduction of the total sleep duration and impair their behavior in the daytime. Several placebo-controlled trials have suggested short-term efficacy of trimeprazine for treatment of night waking in children aged 12 through 24 months (Richman N, 1985) and 12 through 36 months (Simonoff E, Stores G, 1987). Often the infants and young children suffering sleep onset disorders are administered *chloral-hydrate*. First-generation *antihistaminic* drugs possessing hypnotic effect, in particular *diphenhydramine*, are able to reduce sleep latency and number of night awakenings. There are reports on successful use of *clonidine* for sleep disturbances associated with attention deficit hyperactivity disorder (Willens TE et al., 1994). The administration of exogenous *melatonin* has been used in a variety of clinical settings, most frequently in the management of insomnia in the adults. In children, melatonin has been used for chronic insomnia, as well as in the management of sleep disturbances associated with vision disturbances, attention deficit/hyperactivity disorder (ADHD), and neurologic injury (Gordon N, 2000; Griefahn B et al., 2002).

Part 3. Sleep Disordered Breathing in Infants and Young Children

Chapter VIII

Apnea of Prematurity and Infantile Apnea

Apnea of prematurity (AP) is defined as cessation of breathing for periods of 20 seconds or longer, with or without bradycardia or cyanosis, which can result in severe neurological damage and sometimes death. Apnea of prematurity is observed in infants born with gestational age less than 37 weeks (American Academy of Pediatrics, 2003) and may occur in 80% of pre-term infants weighing less than 1000 gm at birth. The probability of an infant to have apnea of prematurity is inversely dependant of the length of gestation, and almost all infants born at gestational age 28 weeks or less suffer AP (National Institute of Health, 1986). About half of the infants born pre-term are characterized by periodic breathing in the neonatal period, and it is common for the apnea of prematurity to be associated with periodic breathing. The American Academy of Pediatrics and the US Birth Cohorts estimate that there are more than 150,000 infants born each year who will be diagnosed and treated for apnea of prematurity. The mechanisms of AP are not clearly defined (Johnson P et al., 1973). Most likely, it is connected with immaturity of the medullary respiratory centers, immaturity of the central and peripheral chemo-receptors and pulmonary reflexes in pre-term babies. This opinion is supported by the fact that AP may be self-limiting phenomenon, likely to disappear with infant's advancing age and maturation. The occurrence of AP will decrease by 80 % when the infant's postconceptional age is 40-52 wk (Albani M et al., 1985). It was also reported that after 43 wk post conception, AP is likely to disappear even in the babies who have been born deeply pre-term (American Academy of Pediatrics, 2003). However, in some cases AP may persist with time, particularly in the infants born at gestational age less than 28 wk (Barrington KJ et al., 1996; Eichenwald EC et al., 1997).

AP are mostly of central origins. Presumably, this is related to so-called oscillatory form of respiration in these infants (Waggener TB et al., 1989). However, many pre-term infants demonstrate obstructive or mixed apneas, suggesting that mechanisms responsible for maintenance of airway patency may be insufficiently developed in this population and be responsible for AP. Reportedly, from 6,5 % to 14 % of the registered AP in pre-term babies were scored as obstructive, and from 31 % to 49,5 % - as mixed (Milner AD et al., 1980; Dransfield DA et al., 1983; Waggener TB et al., 1989; Finer NN et al., 1992; Upton CJ et al., 1992). Complex neuromuscular mechanisms responsible for the maintenance of the UAW patency are often ineffective in functionally immature infants often having numerous co-morbidities (Wilson SL et al., 1980). In particular, anemia, which is common problem in the infants born pre term, may lead to more arrhythmic respiration. Improving hematological findings in these infants may help to control AP. Among the other specific conditions that can provoke episodes of apnea in the infants born pre-term are sepsis, hypoxia, metabolic diseases, cardiac malformations with right-to-left shunt, gastro-esophageal reflux (GER), discoordinated contractions of the pharyngeal muscles, UAW obstruction due to postural changes or due to congenital anomalies (Sheldon HS et al., 1992).

Infants born at term also may suffer pathological sleep apneas known as *infantile apnea (IA)*, or apnea of infancy. Infantile apnea is defined as an unexplained episode of cessation of breathing for 20 seconds or longer, or a shorter respiratory pause associated with bradycardia, cyanosis, pallor, and/or marked hypotonia. This term generally refers to infants with gestational age of 37 weeks or more at the onset of apnea (National Institute of Health, 1986; American Academy of Pediatrics, 2003).

Chapter IX

Obstructive Sleep Apnea Syndrome (OSAS)

Obstructive sleep apnea syndrome (OSAS) in children is a disorder of breathing during sleep characterized by prolonged partial upper airway obstruction and/or intermittent complete obstruction (obstructive apnea) that disrupts normal ventilation during sleep and normal sleep patterns (American Thoracic Society, 1996). OSAS is common in childhood, occurring approximately one-third as often as asthma. OSAS occurs in children of all ages, from neonates to adolescents. It is commonest in the preschool age group, due to adenotonsillar hypertrophy (Jeans WD et al., 1981); younger or older children are more likely to have other underlying etiological factors. The studies have shown similar prevalence rates of approximately 2% (Ali NJ et al., 1993; Gislason T, Benediktsdottir B, 1995; Redline S et al., 1999). In contrast to adults, where the disease occurs primarily in males, in children it appears to occur equally among the sexes (Redline S et al., 1999). This may be because of the lack of hormonal influences in the prepubertal child. Some findings suggest that it is more common in African American children, due to either structural differences or socioeconomic factors (Redline S et al., 1999).

Symptoms include habitual (nightly) snoring (often with intermittent pauses, snorts, or gasps), disturbed sleep, and daytime neurobehavioral problems. Daytime sleepiness may occur but is uncommon in young children. Complications include neuro-cognitive impairment, behavioral problems, failure to thrive, and cor pulmonale, particularly in severe cases. Risk factors include adenotonsillar hypertrophy, obesity, craniofacial anomalies, and neuromuscular disorders. The study was undertaken aimed at evaluation of possible association between maternal-reported snoring and/or noisy breathing in sleep in 2-4-month-old infants and their behavioral characteristics (Kelmanson IA, 2000a). The presence of respiratory symptoms in sleep was associated with infant's mood, in that asymptomatic infants were characterised by more positive mood compared with symptomatic ones, and most negative mood was the feature of those infants who had both snoring and noisy breathing in sleep. This association remained significant after adjustment has been made for major potential confounders. Respiratory troubles in snoring and noisy breathing infants may have real causal effect on infant's behavior, and the same conclusion was drawn from previous interventional studied in children with sleep disordered breathing (Ali NJ et al.,

1996). Causal effect of respiratory disturbances in sleep on negative mood in the infants is substantiated by revealed "dose-response" effect, in that the infants having both snoring and noisy breathing in sleep presented with more negative mood than the infants with only one of the above symptoms. According to previous findings, infants and children whose parents were able to report their snoring had objective evidence for more disrupted sleep (Ali NJ et al., 1993). On its turn, brief sleep was shown to be associated with negative mood in the infants (Weissbluth M, 1981; 1982a). Much less likely seems a reverse causality in that infant's snoring in sleep can upset his/her mother to such a degree that it could in turn affect an infant. Moreover, none of the mothers reported difficulties when handling the infant which is another argument against such an opportunity.

OSAS needs to be distinguished from *primary snoring (PS)* which is defined as snoring without obstructive apnea, frequent arousals from sleep, or gas exchange abnormalities (American Sleep Disorders Association, 1997). Snoring is produced by vibration of the soft palate and faucal pillars and indicates upper-airway narrowing during sleep. Although PS is usually considered benign, this has not been well evaluated, because most studies of snoring children did not discriminate between PS and OSAS. In contrast to OSAS, PS is more common in children; habitual snoring occurs in 3% to 12% of preschool-aged children (Ali NJ et al., 1993; Gislason T, Benediktsdottir B, 1995; Hulcrantz E et al., 1995; Owen G et al., 1995). Table 37 is the summary of the findings on snoring in infants and children coming from different sources. Own findings were that in 129 of 200 cases who entered the study (64.5%), mothers failed to report any symptoms of breathing disturbances in sleep in their infants. Mothers of 10 infants (5.0%) described their babies as habitual snorers. In 48 observations (24.0%), the infants were characterised as usually having other than snoring noisy breathing in sleep. In 13 cases (6.5%), mothers were able to report both snoring and noisy breathing in sleep in their infants (Kelmanson IA, 2000a).

Much interest has been raised in possible association between parental smoking and snoring in children. According to some findings, relationship between parental smoking and snoring in school-aged children with "dose-response" effect was found (Corbo GM et al., 1989). The relative risk of snoring in pre-school aged children associated with maternal smoking was 4,4 (95 % CI = 1,5 - 13,1) (Ali NJ et al., 1993). There no good evidence on similar association in younger children so far, however this issue needs to be further addressed.

In a high proportion of infants and young children who snore in sleep, the underlying cause may be atopy, and the proved allergy to certain allergens has been determined in 36% of these patients (McColley S et al., 1997). The relative risk of snoring associated with atopy was estimated to be 3,9 (95 % CI = 2,0 - 7,7) (Teculescu DB et al., 1992).

OSAS occurs equally among boys and girls (Redline S et al., 1999). The etiology of childhood OSAS is quite different from the adult condition (Table 38). In adults, OSAS is usually associated with obesity. Obese children are also at risk for OSAS, and the degree of OSAS is proportional to the degree of obesity (Marcus CL et al., 1996). However, most children with OSAS are not obese. In fact, they may have failure to thrive. Instead, the vast majority of cases of OSAS in children are associated with adenotonsillar hypertrophy. The peak prevalence of childhood OSAS occurs at 2-8 yr, which is the age when the tonsils and adenoid are the largest in relation to the underlying airway size; endoscopy has shown that the site of collapse is most often at the level of the adenoid (Isono S et al., 1998); and most children with OSAS improve following tonsillectomy and adenoidectomy.

Table 37. Summary of the findings on the frequency of night sleep snoring in children from different age groups

Age group, years	Number observations	Frequency of snoring (95% CI)	Reference
0.5-6	555	3.2% (1,7% - 5,1%)	(Gislason T, Benediktsdottir B, 1995)
0-10	260	11 % (7,8 % - 16,5 %)	(Owen G et al., 1995)
4	325	6,2 % (3,8 % - 9,3 %)	(Hulcrantz E et al., 1995)
4-5	996	12 % (9,7 % - 14,3 %)	(Ali NJ et al., 1993)
5-6	190	10 % (7,8 % - 14,3 %)	(Teculescu DB et al., 1992)
6-13	1615	7,3 % (6 % - 9 %)	(Corbo GM et al., 1989)

Table 38. Obstructive sleep apnea syndrome in children versus adult (Marcus CL, 2001)

Clinical characteristics	Children	Adults
Peak age	Preschoolers	Elderly
Sex ratio	Equal	Predominantly males, Females post-menopause
Etiology	Adenotonsillar hypertrophy	Obesity
Weight	Failure to thrive, normal, obese	Obese
Excessive daytime somnolence	Uncommon	Very common
Neurobehavioral	Hyperactivity, developmental delay	Cognitive impairment, impaired vigilance
Polysomnographic characteristics		
Obstruction	Cyclic obstruction or prolonged obstructive hypoventilation	Cyclic obstruction
Sleep architecture	Normal	Decreased delta and REM sleep
State with OSA	REM	REM or non-REM
Cortical arousal	< 50% of apneas	At termination of each apnea
Treatment		
Surgical	TandA (majority of cases)	UVPP (selected cases)
Medical	CPAP occasionally	CPAP

Notes: CPAP = continuous positive airway pressure; REM = rapid eye movement; TandA = tonsillectomy and adenoidectomy; UVPP = uvulopharyngopalatoplasty.

OSAS also occurs in children with upper airway narrowing due to cranio-facial anomalies observed in cases of *mid-face hypoplasia.* In particular, this may include the infants and children with *Treacher Collins, Pfeiffer, Goldenhar* syndromes. *Mandibular hypoplasia* may also lead to high risk of OSAS. Mild mandibular hypoplasia may manifest as abnormal dental occlusion. However, mandibular hypoplasia can be a component of certain congenital syndromes of multiple inborn malformations, such as *Pierre-Robin syndrome.* On the other hand, micrognatia leading to OSAS may result from perinatal traumas, infections, severe cases of juvenile rheumatoid arthritis. The degree of mandibular hypoplasia clearly correlates with the clinical manifestations of OSAS. Whatever the cause, impaired mandibular growth leads to backward displacement of the tongue, in that the soft palate, posterior pharyngeal wall and the tongue are contacting.

Another reason of OSAS may be neuromuscular abnormalities such as *muscular hypotonia* (e.g., muscular dystrophy (Khan Y, Heckmatt JZ, 1994)) or *muscular incoordination* (e.g., cerebral palsy (Kotagal S et al., 1994)).

Although childhood OSAS is associated with adenotonsillar hypertrophy, it is not caused by large tonsils and adenoid alone. Several facts suggest that it is due to a combination of structural and neuromuscular factors. First, patients with OSAS do not obstruct during wakefulness, showing that structural factors alone cannot be the cause. Second, studies have failed to show a correlation between upper airway or adenotonsillar size and OSAS (Fernbach SK et al., 1983). Third, a small percentage of children with adenotonsillar hypertrophy but no other known risk factors for OSAS are not cured by adenotonsillectomy (Suen JS et al., 1995). Finally, it has been reported that some children were cured of their OSAS by adenotonsillectomy, but developed a recurrence during adolescence (Guilleminault C et al., 1989). Thus, it appears that childhood OSAS is a dynamic process resulting from a combination of structural and neuromotor abnormalities, rather than from structural abnormalities alone. These predisposing factors occur as part of a spectrum: in some children (e.g., those with craniofacial anomalies), structural abnormalities predominate, whereas in others (e.g., those with cerebral palsy), neuromuscular factors predominate. In otherwise healthy children with adenotonsillar hypertrophy, neuromuscular abnormalities are probably subtle. It has been shown that children with OSAS have overall normal ventilatory responses both awake and asleep (Marcus CL et al., 1998b), although subtle differences exist in their response to repeated hypercapnic challenges (Gozal D et al., 1995). It is suspected, however, that children with OSAS may have abnormal centrally mediated activation of their upper airway muscles, leading to a more collapsible upper airway (Marcus CL et al., 1994b).

Children with OSAS appear to have a deficit in arousal mechanisms. Studies have shown that these patients have elevated arousal thresholds in response to hypercapnia (Marcus CL et al., 1998b) and increased upper airway resistance (Marcus CL et al., 1999). Unlike adults, children with OSAS often do not have cortical arousals following obstructive apneas. The findings are that obstructive apneas were associated with arousal in less than half of the apneas in children, and only 18% of apneas in infants (McNamara F et al., 1996). As a result, sleep architecture is more preserved in young children with OSAS (Marcus CL et al., 1994a; Goh DYT, Galster PMCL, 2000), and therefore excessive daytime sleepiness, the cardinal symptom of OSAS in adults, is uncommon (Carroll JL, Loughlin GM, 1995). However, although apnea-related cortical arousals are less common in children than in adults, sub-cortical arousals, as demonstrated by movement (Praud JP et al., 1989; Mograss MA et al., 1994) or autonomic changes (Aljadeff G et al., 1997), occur frequently. It is also possible that subtle disturbances in sleep architecture, which cannot be detected on routine polysomnography, are present (Bandla HPR, Gozal D, 1999). These factors may contribute to neurobehavioral and autonomic complications.

In adults, OSAS is more common during non-REM sleep (Charbonneau M et al., 1994), although some patients do have events primarily during REM sleep. Conversely, in young children OSAS is very much a REM-related disease: a median apnea index may be as high as 56/h in REM sleep versus 9/h in non-REM sleep (Goh DYT, Galster PMCL, 2000). This indicates a state-specific deficit in upper airway/central nervous system function. Furthermore, apneas in infants and young children are longer and more numerous during later REM periods than during REM periods earlier in the night.

Much interest has been raised in possible association between OSAS in infants and young children and influence of certain antenatal factors and child-rearing practices. Own findings confirmed that *maternal smoking during pregnancy* was one of the major risk factors (Kahn A et al., 1994a). PSG-recordings of clinically healthy 509 infants under one year old were the subject of analysis: 115 were undergone PSG study in the neonatal period, and 394 at the age 5 to 29 wk (mean 11 wk) after birth. In 400 observations, mother did not smoke during pregnancy, in 37 cases the mothers were light smokers (smoked 1 to 9 cigarettes daily), and in 72 cases the mothers were heavy smokers (smoked 10 and more cigarettes daily). In the last group, the infants presented with significantly lower birth weight, and more often had profuse sweating during sleep. In comparison with the infants from two other groups, infants whose mothers smoked 10 and more cigarettes daily were characterized by more frequent and longer OA during sleep (Table 39). Relative risk of the infant to have OA in sleep associated with mothers smoking 10 and more cigarettes daily during pregnancy was equal to 2.76 (95% CI = 1.63 – 4.69). Importantly, association between maternal smoking during pregnancy and the number of OA in the infants is characterized by "dose-response" effect, in that the more was the number of cigarettes the mother daily smoked during pregnancy, the more frequent and longer OA were recorded in the infant. When the mother smoked during pregnancy, paternal smoking additionally increased the risk of OA in the infant. Although the mechanism(s) by which maternal smoking during pregnancy can increase the risk of an infant to have breathing disturbances in sleep need to be further clarified, it is worth mentioning that nicotine may interfere with natural maturation of adenilate-cyclase system, lead to changes in the natural course of the fetal brain cells proliferation. Fetal exposure to nicotine is accompanied by disorders of synaptogenesis and production of neurotransmitters, involving such vital sites of the brain as the brainstem (Slotkin TA et al., 1991; Slotkin TA et al., 1992). It was also shown that maternal smoking during pregnancy resulted in increased infant sympathetic tone during sleep (Franco P et al., 1995).

Infant feeding does not markedly influence respiratory rhythm during sleep. Infant sleep position can influence respiratory pattern, however this association depends on infant age and degree of maturity. Thus, in the infants born pre-term and (or) with low birth weight, prone sleep position is accompanied by greater respiratory efficiency, more regular respiratory rhythm, reduced number of apneas, greater value of minute ventilation (Hutchinson AA et al., 1979; Masterson J et al., 1987). However, in the infants aged 3 months, sleep position does nor essentially influence major respiratory characteristics, including the frequency and length of apneas (Orr WC et al., 1985).

Table 39. Statistically significant differences between the infants from the studied groups associated with maternal smoking during pregnancy*

Characteristic	Mother did not smoke during pregnancy	Mother smoked 10 or more cigarettes daily during pregnancy	P
Number observations	400	72	
Maternal vaginal bleedings in pregnancy	24/400	13/72	0.001
Infant birth weigh, g	3360 (900-4540)	3000 (1170-4070)	0.001
Profuse sweating during sleep	14/400	13/72	0.001
Number of OA per 1 hour of sleep	0 (0-1.8)	0.5 (0-1.3)	0.015
Number of infants:			
Who had no OA in sleep	238/400	25/72	0.001
Who had 0.1 and more OA per 1 hour of sleep time	162/400	47/72	0.001
Mean OA length, sec	4.5 (3-14)	7 (3-19)	0.005

* Figures are the absolute numbers or mean values (range)

Untreated OSAS can result in serious morbidity. Although failure to thrive is the exception these days, children with OSAS still tend to have a growth spurt after adenotonsillectomy leading to alleviation of OSAS symptoms (Marcus CL et al., 1994a). Importantly, this does not appear to be due to increased caloric intake postoperatively, but rather to decreased work of breathing (Marcus CL et al., 1994a). An increase in insulin-like growth factor-1 following adenotonsillectomy was found, suggesting that endocrine factors play a role in postoperative catch-up growth (Bar A et al., 1999).

Episodes of upper airway obstruction in OSAS are associated with disorders of hemodynamics and cardiac rhythm (Kelmanson IA, 1997). Initial instability of hemodynamic parameters and cardiac rhythm during sleep becomes most prominent against a background of pathological apneas. Changes in major hemodynamic parameters occurring during and immediately after the episode of OA are presented on the scheme (Figure 10). During apneic episode, blood pressure follows specific dynamics: at the beginning, fast reduction in the values of BP is observed, further the plateau of BP is marked, and when the breathing is resumed, BP quickly increases, not only reaching its initial level, but with some overshot, and it comes to the baseline values only after stabilization of a respiratory rhythm (Schroeder JS et al., 1978; Scharf SM, 1984; Shepard JW, 1985). Fluctuations of BP markedly depend on a degree of blood oxygen desaturation occurring during apnea, and the maximal figures of BP are observed at the moment of maximal desaturation. Systolic BP varies much more than diastolic: while the change in systolic BP may be equal to approximately 2,4 ± 1,3 mm Hg per 1 % of oxygen desaturation, changes in the values of diastolic BP are 1,4 ± 0,8 mm Hg per 1% of oxygen desaturation. It was suggested that certain increase in BP after the termination of apnea is partly connected with increased stroke volume resulted from the increased venous blood return during apnea (due to reduced intra-thoracic pressure). Increased sympathetic tone noticed against a background of apnea may also contribute.

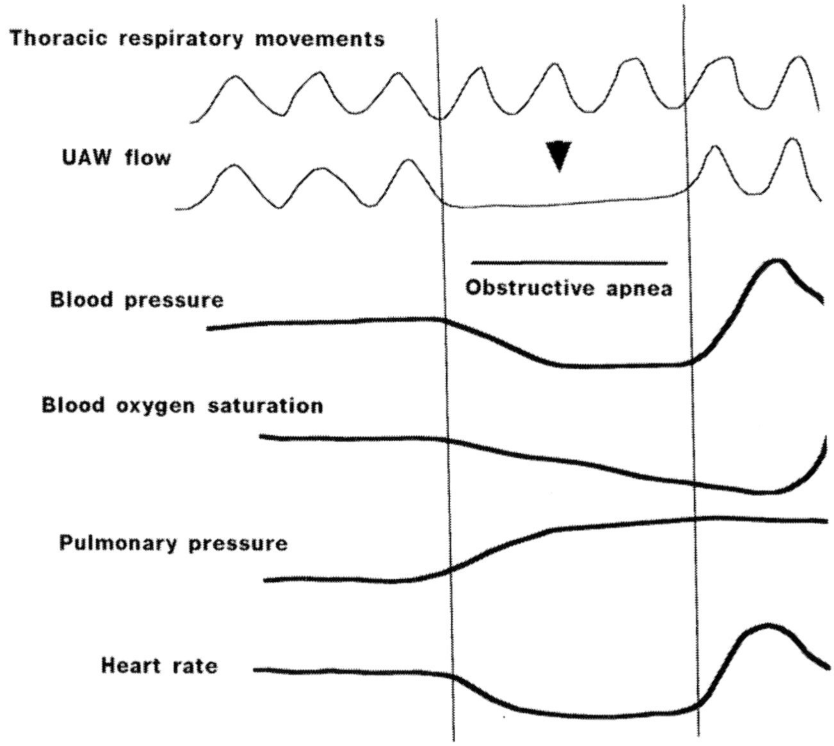

Figure 10. Changes in hemodymanics during obstructive apnea (scheme)

Relationship between pathological sleep apnea and high risk of arterial hypertension has been proved. In particular, pathological sleep apneas were recorded in 30-50 % of the adult patients suffering so-called essential hypertension, and as many as 90 % persons with severe cases of sleep apnea have increased figures of BP (Kales A et al., 1984; Lavie P et al., 1984; Fletcher EC et al., 1985). The system arterial hypertension is a common complication of OSAS in the adults and sometimes may be observed in children (Serratto M et al., 1981). The findings were that the OSAS may lead to increased diastolic BP in children, and such characteristics as apnea index, age of the patient and body-mass index may be of prognostic value (Marcus CL et al., 1998a).

OSAS is accompanied by progressive increase in the pulmonary blood pressure, and there is association between the degree of blood oxygen desaturation during apneas and the degree of increased pulmonary blood pressure (Schroeder JS et al., 1978; Scharf SM, 1984). Similarly to system BP, the maximal increase in the figures of pulmonary blood pressure is observed immediately after apnea, however it does not return to initial level after the end of the episode (Martin RJ, 1987). In the past, cor pulmonale with heart failure was not an uncommon mode of presentation for OSAS in children, but it is now rare, particularly in infancy and young age. Although overt right heart failure now occurs less often, asymptomatic degrees of pulmonary hypertension may be common (Tal A et al., 1988). It has been shown that 37 % children with diagnosed OSAS have reduced right ventricle ejection fraction, although clinical signs of pulmonary hypertension were found only in 7 % cases

(Tal A et al., 1988). When cor pulmonale does develop, it can be reversed by treating the OSAS (Brouillette RT et al., 1982; Tal A et al., 1988).

Pathological sleep apnea may be the cause of cardiac arrhythmias. The most common form is *sinus bradycardia.* This arrhythmia is realized due to vagus-mediated mechanisms connected with modulating influences of arterial chemo-receptors and pulmonary stretch-receptors (de Burgh Daly M, Angell James JE, 1975; Eckberg DL et al., 1985), and this arrhythmia may be alleviated by injection of atropine, and by giving the patient oxygen supply (Tilkian AG et al., 1977; Tilkian AG et al., 1978; Zwillich C et al., 1982). Bradycardia occurring under obstructive events may be considered as protective mechanism aimed at the reduction in myocardial oxygen demands when the blood oxygen saturation is diminished. Sometimes, heart rate may fall to as low figures as 50 beats/minute in the infants and 30 beats/minutes in older children and adults, and the obvious signs of sinus bradycardia are observed in 10-40 % patients with OSAS. Immediately during apneic episode, heart rate is reduced, and after the breathing resumes and compensatory hyperventilation occurs, heart rate will also increase. This dynamics of the heart rate under obstructive event form the picture of *bradycardia-tachycardia.* These episodes of bradycardia-tachycardia occurring during sleep and absent during wake state are characteristic of OSAS.

Increased heart rate variability, normally observed during sleep, becomes particularly noticeable during the episodes of pathological obstructive apneas, and this phenomenon may become so pronounced in children suffering OSAS that can be used as a diagnostic sign. Correlation between respiratory and cardiac rhythms, normally observed in children, remains in patients with sleep apneas during episodes of regular respiration, however it weakens immediately during the episodes of apneas and hypopneas, when cardiac rhythm control is mostly dependent on the mechanisms aimed at the maintenance of metabolic needs (Raschke F, Hildebrandt G, 1987). The features of heart rate fluctuations that may be observed during apneic episodes are their high amplitude and the prevalence of non-respiratory periodic components (Guilleminault C et al., 1984a). Heart rate disturbances accompanying given episode of apnea depend on degree of desaturation and possible co-morbidities (e.g., neuropathies) that can influence balance of sympathetic and parasympathetic tone (Trelease RB et al., 1986). Marked sinus bradycardia is most typical phenomenon in cases of obstructive and mixed apneas, however quite often it may be seen during the episodes of central apneas, especially in young children.

Pathological apneas may be accompanied by episodes of *asystole*, and the frequency with which this form of cardiac arrhythmia may be seen in patients with OSAS ranges from 6 % to 30 %. Likewise, OSAS may lead to the episodes of high degree *AV-block.* Thus, second degree AV-block may be recorded in 4 % to 18 % persons with OSAS (Miller WP, 1982).

Another forms of cardiac arrhythmias associated with OSAS are ventricular and supra-ventricular *premature beats,* as well as various forms of ventricular *tachyarrhythmias* observed at the end of apneic episodes characterized by increased heart rate. Risk of ventricular arrhythmias can be reduced, if the patient with OSAS receives oxygen supplementation, allowing to assume a provocative role of hypoxemia in the origins of heart rate disorders. This trigger mechanism is probably realized through the increased catecholamines production and initial increase in sympathetic tone in cases of obstructive and central apneas, and due to further increase in sympathetic tone in response to hypoxia

(Fletcher EC et al., 1985; Fletcher EC et al., 1987; Somers VK et al., 1989; Ejnell H et al., 1991). It was suggested that cardiac arrhythmias associated with sleep apneas may contribute to the mechanisms of sudden death in some patients with OSAS.

Importantly, the frequency of cardiac arrhythmias in patients with sleep apneas may be diminished, when some therapeutic measures aimed at combating sleep apneas are used, including tracheostomy (Guilleminault C et al., 1983) and CPAP (Becker H et al., 1987). CPAP is also helpful in restoring normal heart rate variability during sleep with predominant respiratory components (Fett I et al., 1991).

Systemic hypertension is a well-described complication of OSAS in adults, and has been reported in a few pediatric case series (Serratto M et al., 1981; Marcus CL et al., 1998a). A systematic study showed elevated diastolic blood pressure in children with OSAS, which could be predicted by apnea index, body mass index, and age (Marcus CL et al., 1998a).

Many reports have suggested that children with OSAS are at risk of neuro-cognitive deficits. Such problems as poor learning, behavioral disturbances, and attention-deficit/hyperactivity disorder have been reported (Weissbluth M et al., 1983; Ali NJ et al., 1994; Chervin RD et al., 1997; Goldstein NA et al., 2000). Snoring in sleep is also associated with certain behavioral peculiarities, including restless sleep, and the relative risk of this form of sleep disorder associated with snoring is equal to 2,1 (95 % CI = 1,5 - 2,8) (Ali NJ et al., 1993). Increased motor activity during sleep in these infants and children was proved by video-monitoring. Similar findings were reported by other authors who have found that relative risk of restless sleep associated with night snoring was equal to 2,6 (95% CI = 1,8 - 3,7) (Gislason T, Benediktsdottir B, 1995). Own findings were that infants with habitual snoring and noisy breathing presented with specific day-time behavioral features, including more negative mood (Kelmanson IA, 2000a). It seems more than likely that this association may reflect more fragmented (Weissbluth M et al., 1984; Ali NJ et al., 1993) and shorter night sleep (Kahn A et al., 1994b), the factors that in turn may lead to more negativism in infants (Weissbluth M, 1981; 1982a).

The natural course and long-term prognosis of childhood OSAS are not known. Specifically, it is not known whether childhood OSAS is a precursor of adult OSAS, or whether these are two diverse diseases affecting discrete populations. While data are lacking on the natural history of the illness, it is possible that these high-risk children will develop a recurrence of OSAS during adulthood if they acquire additional risk factors, such as androgen secretion at puberty, weight gain, or excessive alcohol ingestion (Guilleminault C et al., 1989).

Another phenomenon linked with OSAS in infants and young children is known as *upper airway resistance syndrome* (*UARS),* which was first used to describe a subgroup of patients with conditions that were formerly diagnosed as idiopathic hypersomnia or CNS hypersomnia (Guilleminault C et al., 1993). These terms were used to describe excessive daytime sleepiness without a cause that was clearly defined by a nocturnal PSG or the multiple sleep latency test (Guilleminault C et al., 1992). The patients with UARS displayed repetitive increased upper airway resistance (IUAR) that was defined by increasingly negative inspiratory esophageal pressure and that occurred concomitant with decreased oro-nasal airflow in the absence of frank apnea or oxygen desaturation. These periods of IUAR were brief, typically lasting one to three breaths, and resulted in brief EEG arousals (from 2 to 14 s), followed immediately by decreased upper airway resistance. By definition, UARS patients have daytime sleepiness or fatigue. Children with UARS present with frequent snoring, restlessness during sleep, and sweating (Guilleminault C et al., 1982b). Other characteristics include a change in appetite, poor performance, and problems with behavior (Downey R et al., 1993). Contrary to what is seen in OSAS patients, UARS patients are typically non-obese. They are also frequently younger than OSAS patients. Often, they had mildly abnormal upper airway anatomy (most commonly retrolingual narrowing) as determined by cephalometry. Low soft palates, long uvulas, increased overbites, and high, narrow hard palates have also been described in this syndrome. The factors that mainly determine clinical symptoms are the presence or absence of the full obstruction of the UAW, the efforts needed for a child to maintain breathing during sleep, and the degree of sleep disorders connected with these additional respiratory efforts; less significant seems the presence of apneas confirmed by diagnostic studies (Guilleminault C, 1987). As a rule, children suffering OSAS are characterized by increased UAW resistance (Marcus CL et al., 1999).

Chapter X

Central Hypoventilation

The major causes of central hypoventilation in infants and young children are listed in Table 40.

Congenital Central Hypoventilation Syndrome

Congenital central hypoventilation syndrome (CCHS), also known as the *Ondine curse*, is traditionally defined as the failure of automatic control of breathing (Fleming PJ et al., 1980; Guilleminault C et al., 1982a). Initially, the term "Ondine curse" was coined to describe a syndrome that manifested in adult patients after high cervical/brainstem surgery. When awake and summoned to breathe, these patients did so; however, otherwise, they required mechanical ventilation for severe central apnea. However, soon it has been reported in infants with the typical clinical features corresponding to CCHS (Mellins RB et al., 1970). Although the cases described in adults were markedly different from the typical cases in infants with CCHS, the term Ondine curse gained wide acceptance to denote CCHS in infants and children. It has only been in recent years that the term has fallen out of favor. Since breathing during NREM-sleep is neurologically controlled almost entirely by the automatic system, ventilation is most severely affected during quiet sleep in this disorder (Fleming PJ et al., 1980; Gozal D et al., 1993). However, breathing is also abnormal during active sleep and wakefulness, athough usually to a milder degree. Disordered ventilation control may range from relatively mild hypoventilation during quiet sleep with adequate ventilation during wakefulness to complete apnea during sleep and severe hypoventilation during wakefulness.

Other signs of brainstem dysfunction may be present. It is considered that CCHS is present at birth, even though the diagnosis may be delayed because of variations in the severity of the manifestations or lack of awareness in the medical community, particularly in milder cases. No differences in the occurrence of CCHS are evident on the basis of race. Infants and children of both sexes appear to be affected equally. The exact incidence of CCHS is unknown; however, the estimated incidence is about 1 case per 10,000-15,000 live births, and at least about 200 infants and children with this disease have been described (Idiopathic congenital central hypoventilation syndrome: diagnosis and management, 1999).

Table 40. The major causes of central hypoventilation in infants and children (Marcus CL, 2001)

Primary syndromes	
	Congenital central hypoventilation syndrome (CCHS)
	Late-onset congenital central hypoventilation syndrome
Secondary syndromes	
	Obesity hypoventilation syndrome
	Central hypoventilation associated with brainstem malformations and injuries:
	Arnold –Chiari malformation, type 1 or 2
	Hydrocephaly
	Achondroplasia with foramen magnum stenosis
	Hypoxic-ischemic encephalopathy
	Trauma
	Congenital anomalies (in particular Moebius's syndrome)
	Meningoencephalitis
	Poliomyelitis
	Central hypoventilation associated with other neurologic diseases
	Autonomic dysfunctions (in particular, familial dysautonomia)
	Mitochondrial diseases (in particular, subacute necrotizing encephalomyopathy)
	Neuro-degenerative syndromes
	Other causes
	Drugs
	Hyperthermia
	Hypothyroidism
	Metabolic disorders
	Substances

The exact pathophysiology of CCHS remains unknown and has been the subject of intense speculation and research. It is likely that this is a generalized disorder of the central and/or autonomic nervous system, which affects more than just control of breathing. Genetic defect is the probable underlying etiology for CCHS, especially when associated with *Hirschsprung disease (HD).* Such association has been reported in 15% to 20 % of cases (Haddad GG et al., 1978; Marcus CL et al., 1991b) and was denoted as *Haddad syndrome.* The association of these two relatively rare clinical entities suggests a possible common pathogenetic basis. Another abnormalities associated with CCHS include mediastinal and adrenal ganglioneuromas and other tumors of neural crest origin (Swaminathan S et al., 1989a), as well as ophthalmological abnormalities (Goldberg DS, Ludwig IH, 1996). Mutations in the endothelin signaling pathways have been identified (Weese-Mayer DE et al., 1992). In the absence of HD, the likelihood of a successful search for one or more candidate genes responsible for most cases of CCHS appears relatively unlikely. Conflicting data exist about whether family members of people with CCHS are also affected. On the one hand, family members of people with CCHS do not have evidence of respiratory control dysfunction. On the other hand, many family members have some derangement in autonomic nervous system function.

Multiple attempts have been made to identify structural CNS abnormalities in cases of CCHS. Earlier pathological reports of hypoplasia of the arcuate nucleus in one patient with CCHS and the presence of abnormal evoked potential responses to auditory stimuli further suggested that a brainstem lesion may be present (Paton JY et al., 1989). In addition, central hypoventilation syndrome has been reported in occasional patients with cerebro-vascular malformations or CNS infections. It is likely that more than one brain region is affected in CCHS, and accurate mapping of the several regions involved may provide some clues to the type and function of neurons affected by this condition.

CCHS was thought to be a disorder of central chemo-receptor (hypercapneic) responsiveness (Mellins RB et al., 1970; Shannon DC et al., 1976; Guilleminault C et al., 1982a). Children with CCHS have absent chemo-receptor responses to both hypercapnia and hypoxia, even while awake (Paton JY et al., 1989). However, most children with CCHS arouse to hypercapnia (Marcus CL et al., 1991a), thus the most probable mechanism for CCHS is a brainstem lesion in the area where input from both chemo-receptors is integrated for ventilation (Marcus CL et al., 1991a; Davidson Ward SL et al., 1992a). It has been hypothesized that the ability of CCHS patients to maintain waking ventilation is due to intact chemo-receptor function, which can stimulate the abnormal brainstem centers if the signal is strong enough (Gozal D et al., 1993).

An interesting finding was that body movements during exercise entrained ventilation in the absence of chemo-receptor function (Paton JY et al., 1993). Passive leg motion increase alveolar ventilation in CCHS (Gozal D et al., 1996). The presumed mechanism for this stimulation is by mechanoreceptors in the extremities. Thus, in a setting of deficient integration of respiratory control inputs, mechanoreceptor afferent input from muscles and joints and/or rhythmic entrainment of respiration take over and play a significant role in the modulation of breathing during exercise in children with CCHS.

The beat-to-beat variation of the heart rate was found to be markedly disturbed in patients with CCHS, suggesting widespread dysfunction of autonomic integration (Woo MS et al., 1992). Significant reductions in dopamine turnover as an indicator for catecholaminergic neuronal activity were found in patients with CCHS. Vagally-mediated syncope or asystole may also occur in children with CCHS, thereby lending further support to the assumption that significant dysregulation of central autonomic nervous system control is frequently present in patients with CCHS.

The clinical outcome of children with CCHS has changed markedly since the description of the disorder. In the past, most patients presented with neurocognitive deficits of varying severity (Marcus CL et al., 1991b; Weese-Mayer DE et al., 1992), stunted growth, cor pulmonale, and/or seizure disorders; however, early diagnosis and institution of adequate ventilatory support to prevent recurrent hypoxemic episodes clearly offers the potential for improved growth and development and should be associated with normal longevity. Mortality is primarily due to complications stemming from long-term mechanical ventilation or from the extent of bowel involvement when HD is present.

Late-onset CCHS is defined when the initial signs are found in children aged 2-4 years. It is characterized by respiratory disorders during sleep, as well as numerous hypothalamic disorders, including hypothalamic obesity and other endocrinopathies. In some patients ganglioneuromas, ganglioneuroblastomas may be found (Katz ES et al., 2000).

Secondary Central Hypoventilation Syndromes

Arnold-Chiari malformation is a condition that can affect respiratory control during sleep. In Arnold-Chiari malformation, the hind-brain is elongated, which leads to several associated malformations. Arnold-Chiari type I malformation presents with elongated ("pegged") cerebellar tonsils, displaced into the upper cervical canal. Arnold-Chiari malformation type II is the most common and is usually associated with myelodysplasia and meningomyelocele. Type III contains types I and II and includes cervical spina bifida with herniation of the cerebellum (Nickerson B, 1992). Type IV was also described as cerebellar hypoplasia (Strayer A, 2001), but since it does not involve a herniation of the hindbrain, it is not considered by some to belong in the classification system.

All four types of Arnold-Chiari malformation involve the cerebellum. It was once thought to be a rare finding among very young children (<6 years). However, with the increased use of magnetic resonance imaging, this cranio-cervical malformation is being identified earlier and often in asymptomatic patients before the onset of neurologic symptoms such as headache, neck pain, ataxia, sensory deficit, lower cranial nerve palsies, and nystagmus. Occulomotor disturbances, syncope, torticollis, paralysis, scoliosis, sleep apnea, and sudden death resulting from Arnold-Chiari type I malformation also have been reported (Dure LS et al., 1989; Ali MM et al., 1996; Strayer A, 2001).

Apnea is one symptom in children with Arnold-Chiari malformation type 1 that can be readily assessed. Sleep apnea or central hypoventilation has been reported in a few cases and may be under-recognized and neglected in this population. Arnold-Chiari malformation type I is thought to cause apnea by three possible mechanisms that are related to compression of the brain-stem: depression of the respiratory center, dysfunction of the ascending reticular activating system, or paralysis of movement in the upper respiratory wall including the vocal cord (Shiihara T et al., 1995). As a rule, infant's breathing is not markedly impaired during wake state, manifesting during sleep (Hesz N, Wolraich M, 1985; Oren J et al., 1986). Children with myelomeningocele fail to increase ventilation in response to hypercapnia (Swaminathan S et al., 1989b), arousal reaction in them is also ineffective (Davidson Ward SL et al., 1986). This allows to assume abnormal function of central chemoreceptors or impaired answer of the brainstem structures to the signals from chemoreceptors. In some patients, reduced sensitivity of peripheral chemoreceptors may be found (Swaminathan S et al., 1989b). High risk of associated pulmonary pathology was reported in infants and children with Arnold-Chiari malformation, including restrictive pulmonary diseases resulted from respiratory muscles weakness and (or) scoliosis. As well, pharyngeal dysfunction may be observed due to bulbar paralysis (Putnam PE et al., 1992; Sherman MS et al., 1997). The specified factors increase the risk of aspiration, development of pulmonary athelectases and pneumonias.

Importantly, impaired breathing in infants and children with Arnold-Chiari malformation is not always accompanied by mental disorders.

The outcome in infants suffering myelomeningocele may be serious enough. It was reported that as much as 13 % deaths in children with spina bifida were due to sleep disordered breathing; in 11 % observations the deaths were due to respiratory failure, and in 9 % cases death occurred suddenly during sleep (Kirk VG et al., 1999). Cases of sudden death in asymptomatic children who at the time of autopsy were diagnosed with Arnold-Chiari type 1 warrant greater attention to sleep/respiratory manifestations (Botelho RV et al., 2000).

Prader-Willi syndrome is characterized by obesity, hyperphagia, hypogonadism, mental retardation, hypotonia, and behavioral and sleep disorders. The prevalence of this syndrome is estimated at 1 per 10 000 to 1 per 25 000 life births. The syndrome occurs with equal probability in boys and girls, has no ethnic predisposition. Prader-Willi syndrome is associated with a deletion of the long arm of the paternally derived chromosome 15 (15q11-q13) in approximately 50% to 70% of patients (Mascari MJ et al., 1992). Presumably, primary hypothalamic dysfunction leads to the typical and behavioral manifestations of this syndrome. Signs of autonomic dysfunction were shown in children with Prader-Willi syndrome (DiMario FJ et al., 2000). The syndrome unfolds its typical features by school age, but even in the neonatal period and during the first year of life the child may present with the signs highly suspicious for Prader-Willi syndrome, including marked muscular hypotonia, especially involving neck muscles; poor mimic, weak cry or its absence. Unlike in the adults and older children, appetite may be reduced in the infants, which is often a reason of feeding problems. Typical hyperphagia leading to excessive weight gain may be observed by the age 1-3 years. In the infants and young children, the syndrome may be associated with delayed speech and disorders of articulation.

Patients with Prader-Willi syndrome often present with sleep-disordered breathing. This is characterized by snoring, obstructive apneas, restless movements during sleep, hypoventilation, hypoxia. Typical findings are OA during sleep with desaturation, mostly during REM/active-sleep (Hertz G et al., 1993; Harris JC, Allen RP, 1996). Reaction to hypercapnia is reduced, however this phenomenon markedly depends on the degree of obesity, and with decreasing body weight in a given patient, reaction may be normalized (Arens R et al., 1994). Reduced sensitivity of peripheral chemo-receptors was reported in patients (Gozal D et al., 1994). It was supposed that respiratory disorders in patients with Prader Willi syndrome are associated with impaired central reaction to the impulses arising from peripheral chemo-receptors. It is worth to mention that growth hormone therapy used in patients with Prader Willi syndrome may cause central stimulation of respiratory control, even when the body-mass index remains unchanged (Lindgren AC et al., 1999).

An increased probability of a patient with Prader-Willi syndrome to develop restrictive pulmonary diseases is ascribed to desaturation.

Excessive daytime sleepiness and impaired sleep architecture are also characteristic of Prader-Willi syndrome (Cassidy SB et al., 1990). Sleep often starts with the episode of REM-sleep (which is normally observed only in the infants under one year old).

Achondroplasia is an autosomal disorder characterized by inhibition of endochondral bone formation. Affected children have disproportional shortening of the proximal limbs, a small thoracic cage, and a large head with a depressed nasal bridge. The base of a skull is involved and there is midfacial hypoplasia. Thus, OA is common and may be severe. However, the membraneous bones of the skull grow normally. This results in a large skull resting on a small base with spinal stenosis. As a consequence, compression of the medullary and cervical cord and hydrocephalus may occur leading to impaired central respiratory control and pathological CA and hypoventilation.

Leigh's disease, or subacute necrotizing encephalomyelopathy, is a group of inherited disorders in infancy and childhood characterized by progressive deterioration of brainstem functions (Beckerman RC et al., 1992). Presumably, this is poly-ethiological condition with congenital disorders of metabolism, connected with a deficiency of mitochondrial ATPase. Patients may appear normal in the infancy but develop signs later in life. Clinical findings include feeding problems, vomiting, apnea, alveolar hypoventilation, developmental regress. Sometimes hypoventilation may occur earlier than neurological signs. Brainstem symptoms may include nystagmus, bizarre eye movements, pupilary changes, hypotonia, seizures, sleep/wake disturbances.

There is preferential gray-matter involvement with vascular proliferation, endothelial swelling, progressive neuronal destruction. There is also loss of myelin. Changes are predominantly seen in the midbrain, pons, periaqueductal gray matter, posterior colliculi, medulla, floor of the fourth ventricle, posterior olive (Keens TG, Davidsom Ward SL, 2000).

Joubert syndrome is caused by agenesis of cerebellar vermis with associated episodes of tachypnea (as high as 200 per minute) alternating with prolonged apneas (Harmant van Rijckervorsel G et al., 1983). Patients present with abnormal eye movements, hypotonia and psychomotor retardation. Tachypnea and apnea may be observed both during sleep and wakefulness. Episodes of tachypnea may disappear after the first year of life, however apneas may progress, and impaired ventilation may lead to fatal outcomes.

Obesity hypoventilation syndrome is uncommon in infancy and in young children, however it may become a problem in cases of excessive body weight. Increased adipose tissue around UAW leads to their narrowing (Suratt PM et al., 1983), increased inspiratory pharyngeal resistance (White DP et al., 1985), thus contributing to the origins of OA. However, central respiratory control is also impaired in these cases (Gozal D et al., 1995).

Chapter XI

Apparent Life-Threatening Events (ALTE)

Infants with *Apparent Life-Threatening Event (ALTE)* are presented to medical attention because of an acute and unexpected change in behavior that alarmed the care giver. The initial episodes can occur during sleep, wakefulness, or feeding. They are most commonly described as some combination of apnea, color change (cyanotic or pallid, occasionally erythematous), marked change in muscle tone (limpness, rarely rigidity), choking or gaging (National Institute of Health, 1986). In most cases the observer reported that the episode was potentially life-threatening, and some times that the child had actually died and that a prompt intervention was associated with a normalization of the child's appearance. Episodes of ALTE are most common in infants, particularly under 6 months of age (Wennergren G et al., 1987).

The severity of an ALTE relies only on the history given by the witness. When first examined, an infant may still present with persistent limpness, colour change or abnormal state. There is no possibility to objectively document the severity of the ALTE.

The frequency and the prevalence of ALTE are largely unknown. A study from New Zealand reported a mean annual admission rate for ALTE of 9.4/1000 live births (Mitchell EA, Thompson JMD, 2001). According to Swedish researchers, an incidence of attacks of lifelessness of 0.46 per 1000 live births was found in a prospective study over 24 months (Wennergren G et al., 1987). It has been estimated that 6 per 1000 full-term infants and 86 per 1000 pre-term infants may suffer ALTE (a total of 11 per 1000 live births) (Myerberg DZ et al., 1995). A study from Australia reported that an episode of apnea or cyanosis resulted in hospital admission of 1.37% live births (14/100) (Ponsonby AL et al., 1997).

Episodes of ALTE are more likely to occur in the daytime (Kahn A et al., 1984; Wennergren G et al., 1987). ALTE is a clinical presentation. As such, it requires a systematic and complete evaluation to assess the cause of the event. Only if no medical or surgical cause is found will an ALTE be described as unexplained. Follow-up of the infant will contribute to confirm or disprove the initial clinical classification. The reports of the observers are difficult to appraise because of anxiety and lack of professional experience. Experienced observers can overestimate the severity of the incident as the child's unresponsiveness may be due to a deep sleep state, or underestimate it because of early intervention. However, a detailed history and physical examination are of paramount importance, as invasiveness of the medical evaluation and treatment depends on the index of clinical suspicion.

Questions addressed to the person who witnessed the ALTE aim at a precise description of the event. The following information should be collected in all cases (Kahn A, 2004):

- The events that preceded the ALTE, including recent episodes of fever, illness, vaccination, sleep deprivation, medications, change in daily life routine, feeding habits or life events. Usual sleep conditions should be described, including sleep position, bedding, ambient temperature, use of pacifiers or sedative medications.
- The precise time when the ALTE occurred. Time since last feed should be specified.
- The exact place where the ALTE occurred: the child's cot, parents' bed, a sofa, a car.
- The state of the child: asleep or awake.
- If asleep, the child's body position, the type of bedding, whether the face was covered or free.
- The presence of specific sleep conditions, such as bed-sharing.
- If awake, whether the child was being fed, handled, crying.
- The reason that lead to the discovery of the infant with an ALTE (noise, cry)
- Who witnessed the ALTE.
- The child's appearance when found: consciousness, muscle tone, colour, respiratory efforts, choking, gasping, emesis, sweating, limb or eye movements.
- Any intervention and the child's response.
- The estimated duration of the event.

As in any clinical evaluation, a precise medical history must also be obtained, including pregnancy, birth conditions, neonatal period, medical or surgical problems. Medical treatments and previous evaluations should be described. The family history is completed, including other infants with an ALTE, early deaths, genetic, cardiac or neurologic problems. Parents' smoking and drinking habits should be detailed.

A detailed physical examination should include special attention to neurologic, respiratory, or cardiac abnormalities. General tonus, a careful neurological examination, and a developmental assessment must be done.

Hospitalization is decided on the base of the initial history and physical inspection. It is worth to notice that the episodes of ALTE may be repeatedly observed in 68 % of infants (Ariagno RL et al., 1983), and these repeated episodes commonly occur during first couple of days following the first episode (Wennergren G et al., 1987). Therefore, infant's stay at hospital gives more chances to an infant to undergo proper diagnosis during critical period. However, when the event appears to have been benign and the child normal, the infant may be followed as an outpatient. If the episode appears to be significant with intense stimulation, or when physical examination is not completely normal, the infant should be admitted for further evaluation and placed on continuous cardio-respiratory and/or oxygen saturation monitoring. During hospitalisation, further evaluation are focused on the presenting symptoms and physical examination. ALTE can be a symptom of many disorders. In most studies, over 50 to 70 % of ALTE are explained by some medical or surgical problem (Spitzer AR, Fox WW, 1986; Kahn A et al., 1992b).

There is no uniform accepted protocol of investigations in ALTE cases, enabling to identify underlying problem. Table 41 is the summary of suggested diagnostic studies. If the infant comes for investigation soon after the ALTE episode, there is high diagnostic value of the arterial blood gas analysis (Southall DP et al., 1997).

Major clinical diagnostic groups reported in the literature are listed in table 42, and Table 43 is the summary of the proportion of the ALTE infants related to major diagnostic categories.

Table 41. Suggested list of investigations in infants presenting with ALTE (Kahn A, 2004)

Standard Procedures	Potential diagnosis
Precise child and family history	
Medical examination of the infant	
Scene inspection	Accidental suffocation
Complete blood count, inflammatory tests	Infection, anemia
Blood electrolytes, urea, calcium, phosphate	Dehydratation, hypocalcemia
Blood glucose	Hypoglycemia
Virology and bacterial screening	Infection
Urinalysis and culture	Infection, metabolic screening
Electroencephalogram	Seizure
Chest x-ray	Infection, cardiomegaly
Skull x-ray	Fracture, hypertension
Electrocardiogram	Arrhythmias
Stool culture	Infection
Specific procedures	
Spinal fluid examination	Infection, hemorrhage, metabolic screening
Esophageal pH monitoring	Gastroesophageal reflux
Esophageal barium study (or CT scan)	Anatomical abnormalities
ENT study, laryngoscopy	Airway obstruction or abnormality
Echocardiography	Congenital malformation, cardiac function
Metabolic work-up	Congenital metabolic abnormality
CT scan, MRI	Mass (tumor, hematoma)
Craniofacial study and x-ray	Facial dysmorphia
Intraocular examination	Abuse, shaken infant
Sleep PSG study	Cardio-respiratory, neurological or oxygenation abnormalities
Covert video surveillance	Abuse, Munchausen-by-proxy syndrome

Table 42. Major causes of apparent life-threatening events (Samuels MP, 2000)

Respiratory
Infections (RS-infection, whooping cough, pneumonia)
UAW obstruction (e.g., due to retrognathia)
Lower airway obstruction or closure (e.g., tracheobronchomalacia)
Intrapulmonary shunting (e.g., cyanotic breath-holding)
Neurological
Epileptic - seizure induced
Intra-cranial hemorrhage (vitamin K deficiency, child abuse)
Central hypoventilation - congenital, drugs, neurological diseases
Neuromuscular diseases
Infective
Septicemia, urinary tract infection, gastroenteritis
Meningoencephalitis
Autonomic
Vaso-vagal
Gastroesophageal reflux
Skin color changes
Child abuse
Illness fabrication
Attempted suffocation
Poisonings
Cardiac
Tachyarrhythmias - Wolf-Parkinson-White and long-QT syndromes
Congenital heart disease
Inborn errors of metabolism
Miscellaneous
CO poisonings
Cat smothering
Abnormal infant holding practices
Hemorrhagic shock encephalopathy syndrome
Unknown

Table 43. Frequency distribution of the certain causes of ALTE (Samuels MP, 2000)

Diagnostic category	Percent of ALTE cases related to this category
Respiratory problems	5-25
Epilepsy	7-15
Gastroesophageal reflux	24-62
Induced or fabricated illness	2-16
Metabolic	2-8
Vaso-vagal	11
Unknown	33-55

1. Digestive Problems

About half of all ALTE have been related to digestive problems. Changes in breathing and color have been associated to malformation of the digestive tract, such as volvulus or intussusception; infection; acid gastroesophageal reflux, spontaneous or associated to pyloric stenosis; transit anomalies, such as dumping syndrome, as well as aspiration and choking during feeding or sleep.

Gastroesophageal reflux (GER), defined as the passage of gastric contents into the esophagus, is one of the most common gastrointestinal problems in infants and affects about 50% of healthy, full-term newborns (Jadcherla S, Shaker R, 2001). Postprandial regurgitation is the most common sign of GER in infants and may range from effortless spitting to forceful vomiting. Three mechanisms have been identified that allow reflux of the gastric contents across the esophagogastric junction and into the esophagus: transient lower esophageal sphincter relaxations, hypotensive or incompetent lower esophageal sphincter, and anatomic disruption of the esophagogastric junction. Transient lower esophageal sphincter relaxations are mediated primarily through vagal pathways and account for virtually all reflux episodes in healthy individuals and the majority of reflux episodes (50-88%) in those with GERD. Transient lower esophageal sphincter relaxations are primarily triggered by gastric distension after a meal (Orlando R, 2001). Pharyngeal stimulation, cholecystokinin, and gastric acid secretion may contribute to an increased rate of these transient relaxations. Reflux with normal lower esophageal sphincter pressure may occur with increased abdominal pressure, such as after meals, crying, defecating, and coughing. Infants are more predisposed to "spitting up or vomiting" than are older children and adults because of the small reservoir of their esophagus.

Uncomplicated GER implies a functional or physiological process in a healthy infant with no underlying systemic abnormalities. However, the prevalence of pathological GER is not exactly determined in young children. According to parental reports, 5 to 50% infants under one year had frequent and repeating episodes of regurgitations highly suspicious for pathological GER (Orenstein SR et al., 1996; Vandenplas Y et al., 1996; Nelson SP et al., 1997). Own findings were that frequent, repeated regurgitations highly suggestive of pathological GER were observed in approximately 5 % infants aged 2-4 months (Kelmanson IA, 2000b). Such uncertainty as to the prevalence of GER in infants and young children is due to the fact that the initial diagnosis is mostly based on parental complaints. The parents are not always aware of the symptoms and may overlook this disorder. Therefore, medical personnel should intentionally focus parental attention at possible signs of GER when getting infants history (Orenstein SR et al., 1993; Faubion WA, Jr., Zein NN, 1998). Frequency and severity of GER markedly depend on numerous factors, such as infant's sleep position (Tobin JM et al., 1997; Vandenplas Y et al., 1997), feeding pattern and type of food (Badriul H, Vandenplas Y, 1999). GER is observed with higher probability in the infants born pre-term and/or with low birth weight (Dusick AM, 1997; Ferlauto JJ et al., 1998), low Apgar score (Kelmanson IA, 2000b) and other signs of fetal distress (de Boissieu D et al., 1994; Fonkalsrud EW, Ament ME, 1996). Possible effect of maternal smoking on the risk of GER is still debated (Orenstein SR et al., 1996). Healthy infants may have high prevalence of reflux-associated symptoms, such as daily regurgitation (40%), mild or absent respiratory symptoms, crying more than one hour per day (17%), arching (10%), or daily hiccups (36%) (Orenstein SR et al., 1996). GER that is not complicated by excessive crying, irritability, disturbed sleep, feeding impairment, poor weight gain, or respiratory complications can usually be diagnosed clinically without extensive evaluation. For 85% of infants with uncomplicated GER, the condition is self-limited and usually disappears between ages 6 and 12 months.

Infants with GER are characterized by specific daytime behavioral features. They are more irritable, often cry, have the signs of general distress (Gibson WS, Jr., Cochran W, 1994; Glassman M et al., 1995; Orenstein SR et al., 1996; Catto-Smith AG, 1998). According to own findings, repeated regurgitations in infants were associated with lower infant's distractibility (these babies required more efforts needed to calm him/her) (Kelmanson IA, 2000b). GER may be associated with certain neurological disorders including impaired oral movements, disturbed adaptive reactions (Mathisen B et al., 1999). Changes in the levels of neuro-active peptides, such as beta-endorphines, have been reported (Beyaert C et al., 1995).

Although GER is usually described as feeding-related phenomenon, in certain cases it may be more common finding for infant's sleep, particularly active/REM-sleep, rather than for the wake state (Jeffery HE et al., 1980). Mean duration of GER episodes may increase from 1,5 minutes in a wake state to 5,4 minutes during sleep (Sondheimer JM, 1989). There are numerous sleep-related functional changes that may influence the frequency and duration of the episodes of GER during sleep in infants and young children. Thus, the frequency of swallowing movements playing important role in counteracting GER are markedly reduced during sleep (Bremner RM et al., 1993). As well, the production of saliva necessary for neutralization of the acid gastric contents thrown into the pharynx and esophagus during episodes of GER is also reduced during sleep. The swallows observed in the infants and young children during sleep are "dry", and, as a result, produced esophageal peristaltic wave becomes ineffective, reducing the ability to counteract GER during sleep (Dodds WJ et al., 1973).

GER may lead to disturbed sleep in the infants. It was shown that 75 % of GER episodes in the infants ended up with arousals, accompanied by decreased heart rate and reduced respiratory rate. Sometimes, complete awakenings might occur. Normally, in a short time (on average, during 5 minutes), the infants could fall back to sleep coming back to the same sleep stage that has been observed before the episode of GER (Ramet J, 1994). At the same time, the reports on possible association between GER and arousals are inconsistent. It has been also reported that 41 % of all GER episodes recorded in the infants at the nighttime started and ended up at the moments when the infant has been awake (Kahn A et al., 1991b), and these findings allow to assume that arousals could precede the episodes of GER, rather than follow them. It seems more than likely that GER is one of the signs of autonomous reactions accompanying arousals.

It is important to distinguish GER from *gastroesophageal reflux disease (GERD),* which is a pathologic process in infants associated with poor weight gain, signs of esophagitis, persistent respiratory problems, dysphagia, and changes in neurodevelopmental patterns. GERD occurs in approximately 1 in 300 infants. An infant with GERD is likely to have more than five episodes of reflux per day, regurgitate more than 28 g per episode, refuse feeding, experience episodes of apnea, have problems gaining weight, and demonstrate increased irritability (Reust C, Blake R, 2000). Aspiration pneumonia, chronic cough, wheezing, and other respiratory complications are also associated with GERD. Arching and torticollis (*Sandifer syndrome*) are believed to be related to the esophageal discomfort induced by GERD, as the posturing may help clear esophageal acid (Yellon RF, Goldberg H, 2001). Childhood diagnoses associated with increased risk of GERD include esophageal atresia with repair, neurological impairment and delay, hiatal hernia, bronchopulmonary dysplasia (related to prematurity), asthma, and cystic fibrosis.

In patients with ALTE, the prevalence of recurrent regurgitation or emesis is 60% to 70% (Rosen CL et al., 1983; Tirosh E et al., 1999), and 40% to 80% of patients have abnormal esophageal pH monitoring (Veereman-Wauters G et al., 1991). Case reports have described ALTE triggered by overt regurgitation into the oropharynx or by aspiration of refluxed gastric contents. Gross emesis or oral regurgitation has been correlated with either prolonged apnea (>20 seconds), or with shorter apnea and bradycardia, but the majority of prolonged apnea episodes in these patients were not associated with regurgitation {Tirosh E, Jaffe M, 1996; Tirosh E et al., 1999). Simultaneous recordings of esophageal pH, heart rate, chest wall movement and nasal airflow demonstrated that reflux could precede apnea (Herbst JJ et al., 1979). In selected patients with a history of ALTE, esophageal acid infusion has been shown to induce obstructive apnea (Herbst JJ et al., 1979) or oxygen desaturation (Friesen CA et al., 1994), suggesting that one mechanism by which GER may trigger an ALTE is acid stimulation of laryngeal, pharyngeal, or esophageal chemoreceptors with resultant laryngospasm. When reflux and apnea are associated, the latter are predominantly of the obstructive type (Arad-Cohen N et al., 2000). Despite these reports and the demonstrated potential for GER to cause apnea, several investigations in unselected patients with ALTE have not demonstrated a convincing temporal relationship between esophageal acidification and apnea or bradycadia (Veereman-Wauters G et al., 1991; Kahn A et al., 1992d; Suys B et al., 1994). Although some studies reported an occasional correlation of GER with short mixed or central apneas (5 to 15 sec) (Paton JY et al., 1990), all of the patients reported also had episodes of apnea which were unrelated to episodes of GER, suggesting a primary impairment in the regulation of respiration. The most convincing relationship between GER and episodes of obstructive or mixed apnea has been in infants in whom the episodes occurred while the patient was awake, supine and within one hour of a feeding (Spitzer AR et al., 1984).

Therefore, at present GER should not be considered too easily as a final explanation for the ALTE, and the role of non-acid gastric reflux is still debated.

2. Neurological Problems

Neurological problems account for up to 30 % of explained ALTE. Convulsive disorders, isolated or associated to intracranial hemorrhage, hydrocephalus or developmental deficits were reported. Seizures may require repeated EEG recording before being disclosed. In a young infant, intracranial infection may develop with little symptomatology. Incracranial hypertension may follow the development of undiagnosed brain tumor, subdural hematoma, or a congenital metabolic disease. Major episodes of pallor and hypotonia may follow vaso-vagal reflexes in case of breath-holding spells or other causes of enhanced vagal responses. Rare congenital malformations of the brainstem may be associated with apnea and heart rate abnormalities. Muscular problems leading to hypoventilation and apnea may be associated with various forms of myasthenia. ATLE of neurological origin were described as apneic, choking or rigid, with abnormal body movements. ALTE related to vasovagal responses occurred while the child is awake, often crying.

3. Respiratory Problems

About 20% of explained ALTE result from respiratory problems. These include airway and pulmonary infection such as viral, particularly respiratory syncytial virus, pertussis or mycoplasm infection. Congenital airway abnormalities are reported, including Pierre-Robin syndrome, airway cysts, angiomas, or malacia. Airway obstruction occur with inspiration of vomit or foreign material. Obstructive sleep apnea, are found in association with airway infections, such as CMV or influenza, anatomical or lymphatic tissue abnormalities, hypertrophic tonsils and/or adenoids, as well as allergies or sleep deprivation. Medications, such as phenothiazines or barbiturates favor obstructive sleep apnea in infants. Obstructive sleep apnea can be idiopathic in young infants, and are more frequently found in premature newborns. Obstructions occur at the laryngeal level (Ruggins NR, Milner AD, 1993). Infants of families with multiple histories of ALTE appeared to be more likely to have obstructive sleep apneas than infants of families with only one case of ALTE (McNamara F, Sullinvan CE, 2000). Significant anemia is associated with apnea, in particular in previous pre-term infants. Congenital central alveolar hypoventilation has occasionally been identified following an ALTE. These infants were usually found apneic and flaccid. A history of noisy breathing or snoring and excessive sweating during sleep was reported. ATLE of respiratory origin have a history of apnea, choking, noisy breathing, snoring, and excessive sweating during sleep.

PSG studies could disclose frequent and long apneas in infants admitted for ALTE, both central (Hunt CE et al., 1983), and obstructive (Guilleminault C et al., 1979). Such findings as high UAW resistance (Hartmann H et al., 1998), abnormal periodic breathing (Kelly DH, Shannon DC, 1979), tachycardia and a tachypnea (Oren J et al., 1989), raised arousal threshold (McCulloch K et al., 1982), abnormal reaction to hypercapnia (Anwar M et al., 1993) have also been found. Disorders of sleep architecture were also reported, including altered transition from one sleep stage to another (Haddad GG et al., 1981).

4. Cardiovascular Problems

These conditions account for 5 % of ALTE. Wolff-Parkinson-White, long-QT syndromes, or other heart rhythm abnormalities have been identified. Congenital cardiac malformations, abnormalities of great vessels, myocarditis as well as cardiomyopathy were reported. Some of these infants had a history of episodes of apnea, pallor, cyanosis, limpness, retarded growth. It was also shown that the frequent findings in the infants admitted for ALTE were autonomic dysfunctions, including excessive sweating during sleep (Kahn A et al., 1987), decreased heart-rate variability (Leistner HL et al., 1980), impaired BP control (Fox GPP, Matthews TG, 1989).

5. Metabolic and Endocrine Problems

Congenital metabolic abnormalities account for 2 to 5 % of ALTE. A variety of conditions were reported, including anomalies of mitochondrial fatty acid oxidation, such as medium chain acyl-coA dehyrogenase deficiency (MCADD), leading to sudden and unexpected hypoketotic hypoglycemia, liver failure and acidosis. The disease is associated to various mutations. Urea cycle defects, such as arginase deficiency, lead to acute encephalopathy with brain edema. Fructosemia, galactosemia, Leigh, or Reye syndrome. Neisidioblastosis, Menkes syndrome, hypocalcemia or hypothyroidism have all been found following an ALTE. Most of these conditions are triggered by fasting and are accompanied by vomiting, hypotonia, lethargy and coma. Such diagnosis should be more precisely suspected in case of atypical age for ALTE (before 1 month or over 1 year of life), history of repeated life-threatening events, or infant deaths in the family (Arens R et al., 1993).

6. Other Conditions

Rarer conditions reported in ALTE include nutritional errors, such as excess feeds leading to acute regurgitations. Adverse effects following the administration of drugs have been reported. Accidental smothering and asphyxia in the crib, accidental CO intoxication, drug toxicity or sepsis in non febrile infants were reported. This group of conditions accounts for less than 5% of ALTE. Some ALTE appear to be related to the child's immediate environment. Such condition includes unaccustomed prone sleeping. Similar conditions were reported for infants dying unexpectedly in child care settings (Moon RY et al., 2000). Car safety seats have been associated with apnea and hypoxemia, in premature and in term born infants with irregular breathing (Merchant JR et al., 2001). The role of vaccines in the development of ALTE has been hypothesized. Epidemiological studies lead, however, to the conclusion that a delay in immunization schedule would cause an increase in morbidity and probably, in mortality (Guntheroth WG, 1995). No ALTE was reported following immunization in infants. If from the available literature, it appears that routine immunization is not likely to induce central apnea in normal infants born at term, additional studies are required to assess the safety of new polyvalent vaccines.

The diagnosis of abuse is easy when there are signs of neglect or trauma, such as bruises, burns or bone fractures. Suspicion is raised from a history of frequent relapses of the ALTE, in families with previous histories of apnea of infancy, who have been seeking attention from various health professionals. Recurrent attacks of ALTE that occur only in the presence of the same caregiver, but witnessed shortly by others who are called for help should raise suspiscion of *Munchausen-by-proxy syndrome*. Smothering, imposed suffocation or administration of drugs were reported. It was suggested that less than 3% of ALTE are related to child abuse.

7. Idiopathic ALTE

Up to 50 % of all ALTE remain unexplained despite a highly suggestive history and complete evaluation (Kahn A et al., 1993b; Samuels MP et al., 1993). These were reported as *idiopathic or unexplained ALTE*. The most frequent symptoms reported for this subgroup of infants were apnea, hypotonia, hypothermia, or intense pallor.

Chapter XII

Sleep Disordered Breathing and Chronic Pulmonary Diseases

Obstructive Pulmonary Diseases

The changes of functional respiratory characteristics found in children during sleep can get pathological forms in cases of chronic pulmonary diseases. In patients with initially decreased vital lung capacity, functional pulmonary reserve is reduced, and desaturation may develop more quickly when the respiratory muscular tone and ventilation-perfusion relation during active/REM-sleep are deteriorated. That is why children with normal parameters of oxygenation in a wake state may present with desaturation during sleep, especially in active/REM-sleep (Coffey MJ et al., 1991; Moyer-Mileur LJ et al., 1996). Desaturation during sleep is observed in infants and children with asthma (Chipps BE et al., 1980), cystic fibrosis (Spier S et al., 1984; Coffey MJ et al., 1991), broncho-pulmonary dysplasia (BPD) (Moyer-Mileur LJ et al., 1996), as well as with other chronic obstructive pulmonary diseases (Gaultier C et al., 1985).

Physicians are well aware of the fact that *asthma* is often characterized by severe attacks at the nighttime, both in the adults and in children. Sleep-disordered breathing in these cases is aggravated due to increased bronchial obstruction, reduced lung secretion, suppressed coughing reflex (Bateman JRM et al., 1978), in that tracheal irritation during sleep with greater probability may lead to arousal and (or) apnea, rather than to cough (Sullivan CE et al., 1978). Therefore, in patients with chronic obstructive pulmonary diseases sleep architecture is usually impaired and frequent awakenings may be observed. There are studies on sleep structure in children with asthma (Kales A et al., 1970; Hugues DM et al., 1983). It was shown that in children with asthma who did not stop beta-mimetics and/or euphilline prior to study, sleep architecture was the same as in children from control group (Hugues DM et al., 1983). The same was true for the patients with asthma who discontinued anti-asthmatic medications 12 hours prior to study. Meanwhile, children who have been surveyed within 25 consecutive nights and who stopped anti-asthmatic medications 6 hours prior to the beginning of survey showed marked reduction in proportion of SWS stage 4 without obvious changes in REM-sleep and wake state duration in comparison with children from control group (Tabachnick E et al., 1981). Asthmatic attacks more often occur during active/REM-sleep and in some patients during SWS stage 2, however almost never during delta-sleep (Kales A et al., 1970). Most commonly, asthmatic attacks happen between 02:00 and 07:00. Disordered sleep commonly produces adverse effects on asthma. The role of paradoxical sleep in increased bronchial obstruction is disputable. Studies on experimental animals found increased fluctuations in bronchial and tracheal diameters during REM-sleep, and that has not been observed during SWS. Respiratory characteristics in adult patients with asthma were markedly deteriorated, when measured awakenings in REM-sleep (Shapiro CM et al., 1986).

There are data on functional respiratory parameters in patients with *asthma* obtained by non-invasive diagnostic methods during sleep (Chipps BE et al., 1980; Tabachnick E et al., 1981; Hugues DM et al., 1983). Maximal desaturation is observed in children patients at 3 A.M. REM-sleep in children with asthma is accompanied by reduced respiratory volume, reduced amplitude of thoracic movements, as well as with paradoxical breathing. Blood oxygen saturation level is markedly lower than during SWS. At the same time, diaphragmatic muscular tone in patients with asthma is increased during REM-sleep. The force with which diaphragm contracts during breathing is increased reflecting increased length of a respiratory cycle.

Children with obstructive broncho-pulmonary diseases may present with impaired respiratory control. In particular, severe cases of *asthma* may be associated with disordered function of peripheral chemo-receptors (Kikuchi Y et al., 1994), as well as with reduced sensitivity of respiratory centers to the impulses coming from chemo-receptors (Davenport PW et al., 2000).

The rhythmic nature of *asthma* symptoms, which often peak just prior to the morning awakening, has been widely reported (Bohadana AB et al., 2002; Sutherland ER, 2005). Normal lung function varies throughout the night; it is best around 4 P.M. and worst around 4 A.M. In normal individuals, peak expiratory flow rates and the FEV_1 are greatest in the evening and nadir in the early morning (Goldstein MF et al., 2001; Sutherland ER, 2005). In asthmatics, there is a similar, but exaggerated, pattern with lower flow rates and greater variation in respiratory function (Hetzel MR, Clark TJH, 1980; Goldstein MF et al., 2001; Sutherland ER, 2005). Mean amplitude of circadian variations of total pulmonary resistance was significantly greater in children with stable asthma than in healthy ones (Gaultier CL et al., 1988). Like in the normal subjects, the acrophase (peak value) of pulmonary resistance in children with asthma is noticed at approximately 4 A.M. (Smolensky MH et al., 1981), and this time principally coincides with those periods of sleep when the episodes of dyspnea occur most common in asthmatic patients (Kales A et al., 1970), and when the maximal reduction in blood oxygen saturation may be seen (Chipps BE et al., 1980). This period coincides with the maximum of the registered clusters of REM-sleep on hypnogram, which has been previously discussed. At the same time, the following functional changes that may contribute to increased risk of asthmatic attacks, can be found during specified period: minimal plasma level of catecholamines and cAMP (Reinhardt D et al., 1980), minimum level of catecholamines and corticosteroids secretion (Barnes P et al., 1980), maximal sensitivity to histamine and allergen skin tests, maximal bronchial constriction in response to histamine, acetylcholine and allergens inhalation tests. Cortisol, a natural anti-inflammatory, follows a similar circadian rhythm with peaks in the early morning hours. Cortisol levels are consistently higher in asthmatics (Landstra AM et al., 2002). Of note, this exaggerated peak occurs at the same time as the greatest airway inflammation in asthmatics. Corticotropin levels are also higher in asthmatics, with even greater elevations seen in those with nocturnal asthma (Sutherland ER et al., 2003a). Understandably, the increased cortisol levels reflect increased corticotrophin levels and are likely elevated in response to chronic inflammation. However, this insufficient control of inflammation is likely the etiology for nocturnal symptoms seen in some individuals and may help to explain why patients with nocturnal asthma are more likely to be steroid non-responsive than other asthmatics. Several other phenomena may contribute to increase risk of asthma during specified critical time, including fluctuations in subpopulations of lymphocytes (Levi FA et al., 1985), and antibodies of various classes (Reinberg A et al., 1977). Exhaled nitric oxide, a potent bronchial and vasodilator, is lower in the morning in asthmatics, leading to more bronchoconstriction and an increase in airway resistance (Berry M et al., 2005). This may, in part, be due to or result from chronic inflammation of the airways. Studies using bronchoalveolar lavage and transbronchial biopsies in the lower airways of asthmatics have found an increase in eosinophils and neutrophils, which mediate this inflammatory response. Patients with primarily nocturnal symptoms of their asthma have an even greater number of eosinophils in their airways compared with asthmatics without nocturnal symptoms (Panzer SE et al., 2003). Melatonin also follows a similar circadian rhythm, peaking during the habitual sleep period (Benloucif S et al., 2005). Melatonin has pro-inflammatory effects and is higher in patients with nocturnal asthma, likely adding to the increased symptoms and inflammation at night (Sutherland ER et al., 2003b).

It is also worth to remind that the probability of GER is higher during active/REM-sleep. As high as 55 % of children with *asthma* have the signs of GER confirmed by radiological, fibro-optic studies, and esophageal pH-monitoring (Allen CJ, Newhouse MT, 1984; Nelson HS, 1984; Scheinmann P et al., 1987). There are no specific respiratory signs and symptoms in asthmatic children with GER (Scheinmann P et al., 1987). Although it is not clear whether GER may really serve as a cause of asthmatic attacks in children, it has been reported that acid gastric masses pushed into the distal part of esophagus may provoke bronchial spasm (Davis RS et al., 1983; Baculard A et al., 1986; Herve P et al., 1986), promote increased bronchial obstructive response to inhaled histamine (Wilson NM et al., 1985), to metacholine and in response to isocapnic hyperventilation (Herve P et al., 1986). Presumably, this is due to the vagal reflex caused by irritation of the nervous terminations located in the distal parts of esophagus (Boyle JT et al., 1985). In the case of esophagitis, which is a common complication of GERD, these nerve terminations appear to be even more sensitive to the influences of the acid gastric contents (Andersen LI et al., 1986). Nocturnal GERD can exacerbate asthma symptoms by causing inflammation of both the upper and lower airways. Thus, GER and GERD can probably act as the trigger for bronchial obstruction, and may also increase the effect of other known factors precipitating asthmatic attacks, especially in severe cases (Nelson HS, 1984).

It is important to notice, that exacerbations of asthma and other chronic obstructive pulmonary diseases during sleep may be associated with sleep apneas in these patients. The episodes of OSAS in these cases lead to aggravation of bronchial obstruction episodes. The combination of chronic obstructive pulmonary diseases and OSAS has been referred to as *"overlap syndrome"* (Flenley DC, 1985). The hypoxia/re-oxygenation oxidative stresses in OSAS promote an inflammatory state (Yamauchi M et al., 2005). Respiratory function is lowest in the early morning, the same time that patients with OSAS commonly experience an increased frequency of apneic events. This is due to the greater percentage of REM sleep during this time period. Respiratory function, as previously described, is also lower during REM sleep, with a decrease in both tidal volume and functional residual capacity and more respiratory variation, predisposing to more lability in oxygenation and ventilation. The intermittent hypoxia common to patients with OSAS leads to further decreases in both FEV_1 and vital capacity. Patients with concomitant OSAS and asthma experience lower peak expiratory flow rates.

Increased risk sleep disordered breathing is observed in infants with *broncho-pulmonary dysplasia (BPD)* common complication for infants born pre-term and (or) with low birth weight. BPD is diagnosed in infants born with weight 1500 gm and less who require oxygen supplementation by 28th day after birth. More than 80 % of infants with weight at birth 800 gm and less have the signs of BPD, while in infants with birth weight 1001-1250 gm this complication is observed in 29,7 % cases, and in infants with birth weight 1251-1500 gm the frequency of BPD is 10 % (Fanaroff AA et al., 1995). In some cases, infants with BPD are characterized by normal blood oxygen saturation in the wake state, however desaturation may occur during sleep (Moyer-Mileur LJ et al., 1996). Hypoxemia during sleep may provoke episodes of severe CA and bradycardia in these infants (Garg M et al., 1988). Besides, infants with BPD are characterized by reduced sensitivity of peripheral chemo-receptors (Calder NA et al., 1994; Katz-Salamon M et al., 1995). Frequency and duration of CA in these patients may be markedly reduced by oxygen supplementation (Sekar KC, Duke JC, 1991). Episodes of desaturation in infants with BPD, when SaO2 falls as low as 90% and less, are often recorded during active/REM-sleep. The degree of UAW resistance positively correlates with the number of episodes of marked desaturations during sleep in these babies (Garg M et al., 1988).

Infants with BPD have impaired sleep architecture, particularly diminished proportion of active/REM-sleep and fragmented sleep, and these impairment can be alleviated when the patient is given oxygen (Sher MS et al., 1992; Harris MA, Sullivan CE, 1995).

Sleep disordered breathing is a common finding in infants with *cystic fibrosis.* Maximal reduction in blood oxygen saturation in these patients is found during active/REM-sleep and temporarily coincides with the maximal impairment of ventilation-perfusion relation (Francis PWJ et al., 1980; Muller NL et al., 1980). Transition from quiet to active sleep is accompanied by reduction in inter-costal muscular tone and the tone of diaphragm, as well as by reduced residual pulmonary volume. The subsequent episodes of hypoventilation lead to further reduction in blood oxygen tension. It is noteworthy that the degree of desaturation during sleep in patients with cystic fibrosis does not always depend on degree of clinical symptoms of that disease and the degree with which functional pulmonary disorders are expressed during wake state. This is the argument that these patients need thorough attention during sleep. Therefore, patients with the SaO2 level 95% or less determined during stable state of their disease may need blood oxygen monitoring during sleep, since they are at high risk of hypoxemia during sleep (Standards and indications for cardiopulmonary sleep studies in children, 1996).

Patients with cystic fibrosis are characterized by disorders of sleep quality: reduced sleep efficiency, increased number of sleep stage transitions, increased number of awakenings lasting 5 minutes and longer, longer arousal that may coincide with coughing (Spier S et al., 1984). As such, cystic fibrosis does not lead to increased number of the episodes of pathological apneas in patients (Spier S et al., 1984). At the same time, frequent development of nasal polyps in these patients can contribute to increased probability of OA during sleep (Standards and indications for cardiopulmonary sleep studies in children, 1996).

Restrictive Pulmonary Diseases

There are limited findings on sleep disordered breathing in infants and young children with restrictive pulmonary diseases. However, the findings in adults and older children are the arguments that such association does exist. Relationship between sleep disordered breathing and kypho-scoliosis has been well defined. The vital capacity of the lungs negatively correlates with the degree of spinal deformity in these patients (Canet E, Bureau MA, 1990). As with the other chronic pulmonary diseases, respiratory function is markedly impaired during sleep in these patients. Hypoventilation and desaturation are mostly expressed during REM-sleep, and this is noticed along with the reduced amplitude of thoracic respiratory movements (Sawicka EH, Branthwaite MA, 1987). Maintenance of the adequate respiration during wake state in patients with kypho-scoliosis is mainly reached due to involvement of inter-costal and auxiliary muscles, which become hypo- and atonic during REM-sleep.

Similar findings are true for the patients with inborn skeleton anomalies. *Achondroplasia* is an example. Children with this anomaly are at high risk of sleep disordered breathing: it has been recorded in 48 % of these patients, and in 44 % patients with achondroplasia at least one episode of desaturation with SaO_2 level of 90 % or lower has been recorded (Mogayzel PJ et al., 1988). Apart from compromised ventilatory mechanics, these children are at risk for OSAS due to mid-facial hypoplasia, and for CA and vocal cord paralysis due to foramen magnum stenosis and medullar compression.

Sleep-disordered breathing may result from muscular dystrophy, particularly *Duchenne's muscular dystrophy*. Even before the patients develop obvious respiratory insufficiency due to progressing disease, they may present with hypoventilation during sleep, mostly during REM-sleep, when muscular tone is reduced (Smith PE et al., 1989; Kerr SL, Kohrman MH, 1994; Hukins CA, Hillman DR, 2000). Desaturation during sleep may be found in patients with normal values of oxygen saturation in a wake state. Children with Duchenne's muscular dystrophy have increased frequency of OA during sleep, probably due to UAW muscular hypotonia (Smith PE et al., 1988; Barbe F et al., 1994). Sleep is fragmented in these patients.

Breathing during sleep is also impaired in children with other forms of muscular dystrophies, in particular *spinal muscular atrophy* and *myotonic dystrophy* (Manni R et al., 1993; Labanowski M et al., 1996). In young children, muscular hypotonia and dystrophy of respiratory muscles may lead to increased thoracic compliance (Papastamelos C et al., 1996), contributing to ineffective paradoxical respiration. Patients with spinal muscular atrophy also have specific cranio-facial anomalies, which may add to high risk of OSAS (Houston K et al., 1994).

Chapter XIII

Approaches to Evaluation and Management in Cases of Sleep Disordered Breathing in Infants and Young Children

Apnea of Prematurity, Infantile Apnea, and Pathological Central Apneas

The work-up for infantile apnea and apnea of prematurity begins with a careful history. Factors such as timing of the episode, relationship to feeding, upper respiratory infection, color change, familial history, and degree of resuscitation are critical issues that may enable organization of appropriate studies for evaluating the infant (Spitzer AR, Fox WW, 1986). Particular attention should be paid to possible perinatal complications, Apgar score at birth, length of gestation; it is important to consider infant's body temperature and temperature in infant's ambience, since overheating may often provoke disorder of respiratory rhythm. Data on infant food tolerance should be taken into account. It is also important to specify medications used by lactating mother and/or administered to an infant. As mentioned, alcohol and sedatives may promote respiratory rhythm disorder during sleep, both due to increased risk of central apneas and due to high probability of UAW obstruction. Alcohol and sedatives (mostly, benzodiazepines) potentiate reduction in tone of the muscles participating in the maintenance of the UAW patency and of the muscles of diaphragm (Bonora M et al., 1984; Krol RC et al., 1984). On the contrary, tricyclic antidepressants may selectively activate muscles of the UAW (Bonora M et al., 1985). Much attention should be paid to specific features of infant's behavior: in particular, slackness and increased fatigue during feeding in practically healthy infant may be often indicative for discoordinated respiratory and swallow movements, commonly observed in infants with OSAS (Kahn A et al., 1993b).

PSG-studies not only enable to get objective information on the presence of apneas, but also allow to classify the forms of respuiratory disorders.

Infants presented with episodes of ALTE should be hospitalized and investigated according to the described protocol. Continuous cardiopulmonary monitoring during infant's stay at hospital enables identification of respiratory and heart rhythm disturbances.

Therapy for apnea of prematurity (AP) can be divided arbitrarily into four groupings based on proposed pathogenic mechanisms that might result in apnea. Institution of interventions should occur in the order of increasing invasiveness and risk (Mayock DE, 1997).

1. Increase Afferent Input into the Respiratory Centers

Number and duration of apneas may be reduced by rhythmic sensory stimulation - tactile and/or vestibular. So-called floating or rocking beds are based on this principle (Korner AF et al., 1982; Tuck SJ et al., 1982; Kelmanson IA, 1994); intermittent stimulation turns out to be more effective than continuous one (Saigal S et al., 1986). Various techniques have been implemented: water beds (Jones RAK, 1981; Korner AF et al., 1982) or pneumatic mattresses creating oscillatory movements with periodicity of 12-16 per minute (Saigal S et al., 1986). These devices enable rhythmic raising of head and foot ends of the beds by approximately 3 degrees. Own findings were that during "rocking" the frequency the OA in infants reduced from average index of 1,1 apneas/hours to 0,7 apneas/hours (p=0,003), although the mean duration of apneic episodes did not significantly change (Kelmanson IA, 1994). There are also changes of sleep structure occurring against a background of rhythmic kinesthetic stimulation, including reduced number of transitions from one sleep stage to another, shortening of sleep latency, increased duration of SWS and REM-sleep episodes, reduced infant's activity during sleep and reduced number of muscular twitchings. Traditional cradles may be also successfully used for rhythmic vestibular stimulation (Woodward S et al., 1990). Cutaneous stimulation may be helpful in reducing the risk of AP. Own findings were that massage intervention therapy that included rubbing, stroking, passive movements of the limbs and other means of kinesthetic stimulation in infants born pre-term and/or with low birth weight lead to reduced rate of snoring during sleep which might be a sign of reduction in sleep-disorderd breathing (Adulas EI, Kelmanson IA, 2005).

2. Treatment of Primary Depression of Respiratory Center

a) Treat infection
b) Correct metabolic disturbances
c) Administer central nervous system stimulants (aminophylline, theophylline, caffeine, doxapram).

The exact mechanism by which *methylxanthines* exert their beneficial effect in AP is not known. Proposed mechanisms include increased respiratory drive secondary to increased carbon dioxide sensitivity and increased oxygen consumption. Other mechanisms postulated include adenosine antagonism, enhanced diaphragmatic contractility, and increased cyclic 3',

5' -cyclic AMP. Desaturation spells not associated with apnea are not benefited by methylxanthine therapy. Caffeine is recommended over aminophylline due to it's wider margin of safety and ease of administration (once daily). Table 44 is the summary of recommended indications and regiments of methypxanthine treatment in infants. Apnea, bradycardia, and/or cyanotic spells associated with feeding, handling, suctioning, mucus plugging, etc. should not be counted when determining whether to initiate methylxanthine therapy. Table 45 summarizes pharmacological characteristics of methylxantines most commonly used for the treatment of AP.

Another potent respiratory stimulator is *doxapram* which is used for treatment of AP. This medication presumably influences both peripheral and central chemoreceptors. Doxapram is administered intravevously in preterm infants, when therapeutic dozes of methylxanthines fail to manage AP (Barrington KJ et al., 1987). Doxapram administration prevents the infants from development of severe apneas, leading to decreased PaCO2 level.

If administered in a doze 2 - 2,5 mg/kg /hour, doxapram increases minute ventilation, respiratory volume, and reduces UAW resistance. Doxapram is well tolerated by patients, and side effects are uncommon. Table 46 is the summary of pharmacological properties of doxapram.

Table 44. Guidelines for initiation of methylxantines therapy (Mayock DE, 1997)

Initiation of Methylxanthines		
Apnea Type	***Type of Intervention***	***Treatment Indication***
Spontaneous	No intervention required	Frequent episodes associated with desaturations (SaO2 <80%) and/or bradycardia (HR <90); e.g., one or more per hour over a long period of time such as 12-24 hours
Mild	Light touch, stroke back Associated with desaturations <80% and bradycardia <90	Multiple episodes; more than 6 over a 12 hour period or 12 over a 24 hour period
Moderate	Move infant, i.e. roll over, reposition, etc. Oxygen administered	More than 2 episodes in a 24 hour period
Severe	Prolonged vigorous stimulation PPV with or without oxygen	More than 1 episode in a 24 hour period

3. Treatment of Hypoxemia

a) Treat hyaline-membrane disease, pneumonia, aspiration, etc.
b) Increase inspired oxygen
c) Apply continuous positive airway pressure (CPAP)
d) Treat congestive heart failure
e) Close patent ductus arteriosus
f) Transfuse with packed red blood cells

Table 45. Pharmacological characteristics of methylxantines commonly used in management of apnea of prematurity (Santin RL, Porat R, 2004)

Drug Name	Aminophylline - Stimulates central respiratory drive and peripheral chemoreceptor activity; may increase diaphragmatic contractility. Aminophylline salt is 78.9% theophylline, and theophylline PO is 80% bioavailable; dose adjustment may be necessary when changing from IV aminophylline to PO theophylline. Aminophylline is significantly interconverted to caffeine; if changing from IV to PO aminophylline, increase dosage by 20%. IV and PO aminophylline is effective in 80% of infants with central apnea.
Pediatric Dose	Loading dose: 4-6 mg/kg PO or IV infusion over 30 min Maintenance dose: 1.5-3 mg/kg/dose PO q8-12h or slow IVP q8-12h (start maintenance dose 8-12 h after loading dose) Therapeutic serum concentration (trough) for AOP: 7-12 mcg/mL
Contrain-dications	Documented hypersensitivity; uncontrolled arrhythmias; peptic ulcers; hyperthyroidism; uncontrolled seizure disorders
Interactions	Incompatible with cefotaxime, ceftriaxone, clindamycin, dobutamine, epinephrine, hydralazine, insulin, isoproterenol, methadone, methylprednisolone, penicillin G, and phenytoin; aminoglutethimide, barbiturates, carbamazepine, ketoconazole, loop diuretics, charcoal, hydantoins, phenobarbital, phenytoin, rifampin, isoniazid, and sympathomimetics may decrease effects of theophylline; theophylline effects may increase with allopurinol, beta-blockers, ciprofloxacin, corticosteroids, disulfiram, quinolones, thyroid hormones, ephedrine, carbamazepine, cimetidine, erythromycin, macrolides, propranolol, and interferon
Precautions	Caution in peptic ulcers, hypertension, tachyarrhythmias, hyperthyroidism, and compromised cardiac function; do not inject IV solution >25 mg/min; greater risk of toxicity in pulmonary edema or liver dysfunction because of reduced clearance; monitor heart rate and blood glucose periodically; consider withholding next dose if heart rate >180 bpm; may produce GI irritation, hyperglycemia, CNS irritability, and sleeplessness; may be associated with renal calcifications when used concurrently with furosemide and/or dexamethasone; signs of toxicity include sinus tachycardia, failure to gain weight, vomiting, jitteriness, hyperreflexia, and seizures
Drug Name	Caffeine citrate - Increases respiratory center output, chemoreceptor sensitivity to CO_2, and cardiac output; serum half-life is 40-230 h; half-life is prolonged in cholestatic hepatitis.
Pediatric Dose	Loading dose: 20-40 mg/kg PO or IV over 30 min; equivalent to 10-20 mg/kg of caffeine base Maintenance dose: 5-8 mg/kg PO qd or slow IVP; equivalent to 2.5-4 mg/kg of caffeine base; start maintenance dose 24 h after loading dose
Contrain-dications	Documented hypersensitivity; products containing sodium benzoate
Interactions	Cimetidine may impair caffeine hepatic metabolism, increasing clearance and half-life; phenytoin induces hepatic metabolism of caffeine, decreasing its half-life and increasing clearance; increases metabolism of phenobarbital; increases its own metabolism; can reduce theophylline clearance by 25% and increase elimination half-life; may enhance cardiac inotropic effects of beta-adrenergic stimulating agents; slightly increases urine levels of VMA, catecholamines, and 5-hydroxyindoleacetic acid
Precautions	Restlessness, vomiting, and functional cardiac symptoms (eg, tachycardia, extrasystoles, palpitations) possible

4. Avoidance of Triggering Reflexes

a) Beware of suction catheters
b) Avoid nipple feedings (feed by tube or intravenously)
c) Avoid hyperinflation and hyperventilation during bagging
d) Avoid cold stimuli to the face
e) Avoid severe flexion of neck
f) Treat gastroesophageal reflux

Carnitine supplementation has been used for both prevention and treatment of apnea in premature infants, however there are no strong evidences as to the real benefit of this therapy so far (Kumar M, 2005).

The diagnosis a *central hypoventilation syndrome (CHVS)* in infants and children is usually based on clinical and PSG findings. Infants with CCHS may have significant hypotonia and temporary feeding difficulties. In addition, moderate-to-severe gastroesophageal reflux is frequently present. Early administration of prokinetic agents and antireflux medications in patients with hypotonia, temporary feeding difficulties, and gastroesophageal reflux is often required; nasogastric feeding to provide adequate nutrition may be needed. More radical approaches, eg, percutaneous gastrostomy tube feeding insertion and/or antireflux surgical procedures, may be necessary if these problems are severe or persistent. For infants with HD, surgical intervention and, sometimes, colostomy to relieve the distal intestinal obstruction may be required. Although little attention has been given to the visual system of children with CCHS, ophthalmoplegia and other ocular anomalies have long been recognized to be occasionally present; therefore, a thorough and periodic ophthalmologic evaluation is necessary.

Table 46. Pharmacological characteristics of Doxapram (Santin RL, Porat R, 2004)

Drug Name	Doxapram - Stimulates respiratory drive by activating peripheral carotid chemoreceptors; increases tidal volume and slightly increases respiratory rate; stimulates medullary respiratory center with increasing doses; induces pressor response due to improved cardiac output (greater in hypovolemic patients).
Pediatric Dose	0.5-2.5 mg/kg/h continuous IV infusion
Contraindications	Documented hypersensitivity; epilepsy; mechanical obstruction; muscle paresis; flail chest, pneumothorax, other restrictive lung diseases; head injury; stroke; significant cardiovascular impairment; severe hypertension; intraventricular hemorrhage and kernicterus reported when used in a neonate's first week of life or when the bilirubin level is high.
Interactions	Delay infusion for at least 10 min after discontinuation of anesthetics known to cause myocardium sensitivity to catecholamines; may temporarily mask residual effects of muscle relaxants; additive pressor effect may occur if coadministered with MAOIs
Precautions	Caution in neonates; contains benzyl alcohol, which has been associated with a fatal gasping syndrome in premature neonates; incidence of hypertension and CNS symptoms increase at infusion rates >1.5 mg/kg/h

Attempts to enhance the respiratory stability and promote eucapnia in patients with CCHS failed when pharmacologic approaches were employed. Trials with doxapram did not show consistent improvements in spontaneous ventilatory or gas exchange parameters.

Most centers that provide long-term home care for children with CCHS employ positive pressure ventilation through a permanent tracheotomy (Villa MP et al., 1997). The types of positive pressure ventilators used in the home vary among centers and have gradually evolved, reflecting the dynamic needs of the population of patients with CCHS and the technical developments in the field. Experience with negative pressure ventilation in patients with CCHS has been quite favorable (Marcus CL et al., 1991b). This ventilatory modality is cumbersome and requires significant equipment adjustments over time. In addition, noninvasive ventilatory support may be associated with upper airway obstruction in younger children with CCHS, most probably elicited by absent pharyngeal dilator activation during non–centrally mediated inspiration. Transition from invasive mechanical ventilation to nasal mask ventilation has been reported in patients with CCHS who are older than 7-8 years and who were nocturnally dependent on the ventilator. This transition generally fails in patients younger than 5 years or in children who require ventilatory support during the daytime; however, recent reports of infants receiving adequate noninvasive ventilatory support for more than 2 years suggest that nasal mask ventilation may be a viable option in selected patients, even at younger ages.

Daytime *diaphragm pacing* in children with CCHS provides greater mobility than mechanical ventilation (Weese-Mayer DE et al., 1989; Dobelle WH, 1994). Thus, candidates for diaphragm pacing are potentially ambulatory patients who require ventilatory support 24 h/d via tracheotomy and who do not exhibit significant ventilator-related lung damage. Diaphragm pacer settings must provide adequate alveolar ventilation and oxygenation during rest and daily activities. Major disadvantages of diaphragm pacing include cost, discomfort associated with surgical implantation, and potential need for repeated surgical revisions because of pacer malfunction. Despite these potential limitations, most parental reports regarding diaphragm pacing are favorable. The recent development of a quadripolar electrode offers several advantages that primarily include greater durations of diaphragmatic pacer support at diminished risk of phrenic nerve damage, diaphragmatic fatigue, and optimization of pacing requirements during exercise.

Surgical interventions include traditional procedures. *Tracheotomy* may be indicated for ventilatory support. Colostomy is sometimes required when HD is present. When feeding problems arise, particularly during infancy, gastrostomy tube placement with or without antireflux procedures may be required. Usual postoperative follow-up care for these procedures is necessary but does not differ from the care needed by any other patient.

Obstructive Sleep Apnea Syndrome

Snoring during sleep may be highly indicative of OSAS in infants and young children, although parents often do not pay much attention to this symptom and wrongly believe that this is a harmless phenomenon. Therefore, physicians should specifically address this symptom when getting information on clinical history. Increased daytime sleepiness, infant's behavioral disorders including increased irritability, low distractability may be indirect evidence of sleep disorderd breathing. It has been proposed to calculate *clinical index of obstructive apneas,* which enables to reliably identify patients with OSAS (Brouillette RT et al., 1984). Following variables are considered when calculating this index: impaired respiration, apneas and snoring during sleep.

Impaired respiration is estimated using 4-grade scale:

0 – never observed,
1 – observed sometimes,
2 – observed often,
3 – observed always.

Presence of apneas is estimated on two-point scale: 0 - no, 1 - yes.

Snoring is considered on a 4-point scale:

0 – never observed,
1 – observed sometimes,
2 – observed often,
3 – observed always.

The following formula for the estimation of the probability of OSAS in a child was offered:

1.42 x (impaired respiration) + 1.41 x (apneas) + 0.71 x (snoring) – 3.83

The score more than 3,5 corresponds to high probability OSAS, often requiring adenotonsillectomy, and the score less than -1 is practically never associated with OSAS. Values from -1 to 3,5 should be considered as indications for PSG-study.

Physical inspection of and infant and child should assume examination of respiratory and cardiovascular systems. Since anatomic changes of the UAW may contribute to the origins of OA, these changes, including minor anomalies, need to be excluded. The CT scans of the head and the neck performed in some patients suspicious for OSAS were helpful in finding out that even in a wake state oropharyngeal lumen in these subjects was narrower than in the normal ones (Bohlman ME et al., 1983; Haponik EF et al., 1983; Suratt PM et al., 1983). Ultrasound investigation is also used to study the structure of pharynx, larynx and trachea. Craniofacial anomalies, even minimally expressed, may be also associated with high risk of UAW obstruction and development of OSAS in children.

Excessive daytime sleepiness may be confirmed by the means of *multiple sleep latency test (MSLT)* (Guilleminault C, 1987). This test means that sleep latency is repeatedly determined during 24-hour time period, usually with 2-hour intervals. Meanwhile, this study cannot be readily performed in infants and very young children, and there are no reliable normative values of this test in the specified age group.

Attempts were made to use pharmacological approaches for management of infants with OA. Along with methylxantines and doxapram that were mentioned above as the tools of the management of CA and that are also used in the cases of OA in infants, such medications as anticulvulsants (Kelly DH, Shannon DC, 1981) and atropine deriviatives (Kahn A et al.,

1991a) have been applied. However, their benefit and indications for use need to be further studied.

Most children with OSAS markedly benefit from *adenotonsillectomy* (Suen JS et al., 1995). Although these children are improved, it remains unclear, whether or not they are entirely cured. In post-operative period, children with OSAS who underwent adenotonsillectomy are at risk of respiratory disorders due to airway congestion, increased secretion, depressed respiration due to use of analgesics and local anesthetics, as well as due to pulmonary congestion (Galvis AJ, 1987). Respiratory disorders in postoperative period have been described in 16-27 % of children with OSAS (McColley SA et al., 1992; Rosen GM et al., 1994; Ruboyianes JM, Cruz RM, 1996). Therefore, young children with OSAS should be undergone adenotonsillectomy at hospital, and ideally they should be subjects to overnight PSG-study during next 6-8 weeks after operation.

Continuous positive aiway pressure (CPAP) therapy via nasal mask is used in infants and children with OSAS (Remmers JE et al., 1984). CPAP increases intraluminal upper airway pressure and elevates it above the upper airway critical closing pressure, thereby stenting UAW open. It reduces UAW and diaphragmatic inspiratory muscle activity. Muscle assuaring UAW potency are activated by CPAP (Mathew OP et al., 1982). Creation opositive aiway pressure may be used with diagnostic purposes to define closing pressure, i.e. the minimal value of sub-atmospheric pressure required for UAW obstruction. This closing pressure varies and depends on a sleep stage: it is higher during REM-sleep and stages 1 and 2 of quiet/NREM-sleep, than in SWS (stages 3 and 4 of quiet/NREM-sleep) (Issa FG, Sullivan CE, 1984).

Though use of CPAP technique is not usually recommended to patients with body weight less than 30 kg, there are data on successful use of this procedure and on its good tolerability in infants and young children (Guilleminault C et al., 1995; Marcus CL et al., 1995b). There is always a risk that increased expiratory airway pressure due to CPAP can provoke central apneas and central hypoventilation in some inafnts (Waters KA et al., 1995). This effect may be connected with activated Hering-Breuer reflex. At the same time, there are findings that PaO2 values were higher in children with pathological CA who received CPAP therapy (Miller MJ et al., 1985).

Supplemental oxygen therapy has been considered as an option in chidren with OSAS. This therapy resulted in improved blood oxygenation during sleep in these patients (Marcus CL et al., 1995a; Aljadeff G et al., 1996). At the same time, there are no clear pathophysiological grounds for application of this method in cases OSAS: increased blood oxygen tension does not markedly influence arousal threshold during sleep, neither does it reduce the work connected with the maintenance of respiration in cases of OA. Supplemental oxygen is justified in the newborn OSAS patients with minimal cranio-facial anomalies, when one can expect that these amonalies may be alleviated with infant's further growth and development. As well, supplemental oxygen may be beneficial in patients with OSAS, when adenotonsillectomy ineffective, CPAP is not well tolerated, and tracheostomy cannot be performed for certain reasons.

Uvulopalatopharyngoplasty (UPPP) is a surgical method in combating with OSAS (Fujita S et al., 1981). This surgical operation is aimed at remodelling configuration of the UAW. Operation includes resection of tonsils, uvula, part of the soft palate and additional mucosal folds on the back wall of the oropharynx. In some cases, UPPP is performed in combination with operations on hypoglossal bone and on the mandible (Powell N et al., 1983; Riley R et al., 1984). Pediatric patients for whom this procedure seems to be most recommended are the children with craniofacial anomalies or anomalies of the neck. They may require frequent repeated inspections, including PSG-studies, to make a decision as to the real need in surgical interventions. UPPP has been reported to be successful in children with cerebral palsy and hypotonic upper airway muscles (Kosko JR, Derkay CS, 1995), however, it has not been extensively studied in the uncomplicated pediatric patient.

Craniofacial surgery is appropriate for some children with craniofacial anomalies (Burstein FD et al., 1995; Schmitz JP et al., 1996).

The use of *oral appliances has* not been reported in infants and children, partly due to the concern that the appliances may adversely affect the facial configuration of the growing child.

Tracheotomy is an effective method of treatment in the heavy, complicated cases an OSAS when tonsillectomy or adenotomy appear insufficient, and UPPP cannot be performed for technical reasons or inefficient. Tracheotomy invariably leads to disappearance of obstructive apneas (Guilleminault C, 1987), however, as a heavy psychological trauma, it should be avoided until the other methods of treatment fail to solve the problem.

Sleep Disordered Breathing in Chronic Pulmonary Diseases

Good supportive therapy is essential in patients with chronic lung disease and nocturnal hypoxemia. Medical management should be maximized with bronchodilators (including long-acting agents, when appropriate), antiinflammatory agents, diuretics, etc. Nutrition should be optimized. In patients with ventilatory muscle weakness, chest physical therapy and devices to enhance coughing and secretion clearance are useful. Prevention of aspiration (in patients with bulbar muscle weakness) and treatment of scoliosis will help prevent further deterioration in lung function.

Supplemental oxygen is usually given even for mild nocturnal hypoxemia in infants and young children, although strict guidelines have not been developed (Poets CF, 1998). It is thought that adequate oxygenation is necessary for neurocognitive development, cardiac function, and optimal growth. Correction of sleep-related hypoxemia has been demonstrated to improve growth in infants with BPD if SaO2 is maintained at level at least 93%. Supplemental oxygen is used frequently in patients with cystic fibrosis (Schidlow DV et al., 1993). It has not been shown to improve sleep quality.)

The use of *noninvasive positive pressure ventilation (NIPPV)* for both acute and chronic respiratory failure in children has been increasing rapidly in recent years. NIPPV is being used frequently in children with advanced cystic fibrosis, either as supportive therapy or as a bridge to transplantation (Hodson ME et al., 1991; Padman R et al., 1994). Studies have shown an improvement in oxygenation, ventilation, and vital capacity, as well as subjective symptomatic improvement. NIPPV via nasal mask has been used widely in children with ventilatory muscle weakness, including Duchenne muscular dystrophy and spinal muscular atrophy (Padman R et al., 1994). Negative pressure ventilation is less useful, as patients may develop a worsening of upper airway obstruction (Hill NS et al., 1992). Patients with ventilatory muscle weakness and chronic respiratory failure, treated by nocturnal NIPPV, improve gas exchange during spontaneous breathing while awake. NIPPV is thought to rest the respiratory muscles, and adequate nocturnal ventilation may result in resetting of the central ventilatory drive. In adults with muscular dystrophy, sleep efficiency and sleep architecture improved on NIPPV (Barbe F et al., 1996). Patients receiving NIPPV have fewer hospitalizations (Padman R et al., 1994). Side effects and complications of NIPPV are similar to those of CPAP. Some children with ventilatory muscle weakness will ultimately require a tracheostomy and positive pressure ventilator for secretion clearance, if other mucus-clearing techniques are unsuccessful. In addition, ventilation via tracheostomy is the most practical mode of ventilation in children requiring ventilatory support 24 h/d, in order to facilitate eating and speech.

Nocturnal *asthma* may require administration of long-acting methylxantines, enabling their peak blood level during night hours. From the chronobiological point of view, it is unreasonable to maintaine stable blood concentration of methylxantines, since the maximal clinical benefit of these medications is noticeable at night, when the patients are at maximum risk of asthamtic attacks (Smolensky MH et al., 1981).

Reflux-Associated Apnea

Gastroesophageal reflux is a common findling in apneic infants and children. Although in certain cases causal relationship between GER and apneic spells may be disputable, this underlying condition should be duly identified and treated to avoid possible complications.

Thorough history and physical examination with attention to warning signals is generally sufficient to allow the clinician to establish a diagnosis of GER, recognize complications, and initiate management (Rudolph CD et al., 2001). A thorough prenatal and neonatal history should be obtained including complications with pregnancy or birth, prematurity, newborn screening, growth and development, recurrent illness, surgery, and hospitalization. A family history of significant illness, gastrointestinal or metabolic disorders, and allergy should be documented. In the infant with suspected GER, the health care provider should observe the caretaker feeding the infant. This allows the provider to assess feeding technique, swallow function, as well as infant-caregiver interaction or bonding. The infant's height, weight, and head circumference should be documented at each visit. Significant GER occurring in the absence of vomiting may manifest as failure to thrive. Chronic acid reflux into the oral, pharyngeal, and nasal cavities may result in erosion of mucous membranes and dental enamel. Improper oral hygiene techniques may also result in tooth decay, so parents should be questioned about care of their infant's teeth.

Infants with severe forms of GER and with the signs and symptoms suspicious for GERD, such as poor weight gain, esophagitis (hematemesis), or respiratory symptoms require a referral to a pediatric gastroenterologist for further diagnostic tests. Diagnostic tests commonly used to determine the presence of GERD include the upper gastrointestinal (UGI) series, esophageal pH monitoring, esophageal endoscopy and biopsy, and nuclear scintiscan.

The barium contrast UGI series is important, particularly in infants, to identify anatomic causes of recurrent regurgitation or vomiting, such as pyloric stenosis, malrotation, hiatal hernia, esophageal stricture, gastric outlet obstruction, or other anatomic abnormalities. The presence of reflux and delayed gastric emptying time may also be identified. However, the UGI series is not sufficiently sensitive or specific to provide a definitive diagnosis of GER (Rudolph CD et al., 2001). *The "gold standard" for the diagnosis of GERD is the 24-hour pH probe*. The test is performed by transnasally placing a microelectrode into the lower esophagus, which measure and records intraesophageal pH usually every 4 to 8 seconds. Esophageal pH monitoring via a 24-hour probe provides information about the rate of gastric clearance and frequency of reflux. Endoscopy involves both visualization and biopsy of the esophageal epithelium. Nuclear scintiscan is performed by oral ingestion of radionuclide-containing formula or food. The stomach, esophagus, and lungs are then scanned for evidence of GER and aspiration.

Conservative management of the thriving infant with uncomplicated GER is indicated if there are no complications. Non- pharmacological interventions are generally all that are required, and UGI series is not required unless there are signs of gastrointestinal obstruction (Rudolph CD et al., 2001). Other diagnostic tests may be indicated if there are signs of poor weight gain, excessive crying or irritability, sleep disturbances, or respiratory problems. Life style changes include the following modifications (Arguin AL, Swartz MK, 2004):

(1) Feeding schedule;
(2) Thickening of feedings;
(3) Positioning;
(4) Formula changes.

In the overfed infant or infants fed large volumes at infrequent intervals, the volume of formula should be reduced.

Thickening feeds with rice cereal is associated with a decrease in the number of vomiting episodes, but does not improve reflux index scores (Rudolph CD et al., 2001). Thickened formula may reduce the frequency of regurgitation and total volume of emesis. It may also reduce time spent crying and increase time spent sleeping. Because rice cereal- thickened feedings have 150% of the caloric density of un-thickened feedings, only 65% of the volume needs to be fed per meal; thereby decreasing gastric volume (Orenstein S, 2001). In an infant who is solely breast-fed, it is not possible to thicken the feeds unless the mother uses a breast-pump and then feeds the infant, which can be tedious and less desirable. Commercially available pre-thickened formulas which contains added rice, may be advantageous because they are nutritionally balanced, convenient, well tolerated, and the viscosity is two-to-threefold less than when rice cereal is added to formula. They can be given without an enlarged nipple hole.

Holding the infant in a head-elevated position for 20-30 minutes after feeding may reduce the signs of GER (Arguin AL, Swartz MK, 2004). The prone position has also been shown to reduce reflux, aspiration, and crying time and speed gastric emptying. However, parents should be cautioned that *placing the infant in a prone position should only be done when the child is awake and can be continuously observed. Prone positioning during sleep may be only considered in unusual cases of GER where the risk of death from complications of GER outweighs the potential increased risk of SIDS associated with infant's prone positioning in sleep* (this issue is discussed in the next book section) (Rudolph CD et al., 2001). The semi-supine position after feeding, such as when placing the baby in an infant car seat, exacerbates GER and should be avoided (Rudolph CD et al., 2001).

Many parents and families attribute GER to sensitivities or allergies to milk or formula. True allergy to cow's milk or soy-based formulas occurs in less than 2% of infants and is unlikely to present with vomiting as the only symptom. However, in formula-fed infants, it is reasonable to try a protein hydrolysate or elemental formula for a 2-week period if a cow's milk protein intolerance is suspected (Rudolph CD et al., 2001). In an infant who is breastfed, temporary elimination of cow's milk protein from the mother's diet may be tried.

In the infant with severe GER and suspected GERD with recurrent vomiting and irritability, in which other causes of vomiting have been excluded, two different approaches may be taken (Arguin AL, Swartz MK, 2004):

(1) empiric treatment with either a sequential or simultaneous 2-week trial of hypoallergenic formula and acid suppression may be initiated. If there is improvement, pH monitoring to determine adequacy of therapy or upper endoscopy and biopsy to diagnose esophagitis may be performed;
(2) alternatively, evaluation can begin with esophageal pH monitoring to determine if episodes of irritability and sleep disturbance are temporally associated with acid reflux.

In infants and toddlers with chronic vomiting or regurgitation and recurrent episodes of cough and wheezing, a 3-month trial of acid suppression therapy of GER is recommended in addition to life style changes. A trial of prolonged medical therapy for GER is recommended if esophageal pH monitoring reveals increased frequency or duration of esophageal acid exposure (Rudolph CD et al., 2001). Persistent problems and complications of GER, despite non-pharmacologic interventions, may also warrant pharmacological intervention. Acid suppressants are generally the first line of therapy. This group of medications includes *histamine-2 receptor antagonists* (H-2 blockers), *proton pump inhibitors*, and *antacids* (Table 47). The use of antacids in pediatric patients is generally not recommended, since significant aluminum absorption from antacids can occur approaching the levels reported to cause osteopenia and neurotoxicity. Prokinetic agents such as *metoclopramide* may improve gastric emptying and esophageal motor tone. *Surgery* may be indicated if reflux causes persistent vomiting with failure to thrive, chronic esophagitis, esophageal strictures, or chronic pulmonary disease that does not respond to 2 to 3 months of pharmacological therapy.

Respiratory Monitoring

Much interest has been raised in the benefits of *respiratory monitoring* in infants and young children with sleep disordered breathing. In a simplest case, respiratory monitors are the portable, user-friendly devices that can sence thoracic respiratory movements and signal about respiratory pauses of a definite duration set in advance (National Institute of Health, 1986; Spitzer AR, Fox WW, 1986; Grögaard JB, 1993). Some monitors are designed to allow simulataneous recording of respiratory movements and heart rate. However, these monitors are mainly aimed at registration of CA, and may turn out to be inefficient in cases of OA, when thoracic respiratory movements do not cease. None of the current monitors available for home use will detect OA reliably (American Academy of Pediatrics, 2003).

Valuable information comes from so-called *multichannel recordings*. These recordings are microprocessor-based systems that measure chest wall movements, nasal/oral airflow, respiratory paradox, saturation, heart rate and body movements, sometimes snoring sound. EEG and EMG are not typically recorded in most portable systems, and sleep staging may not be possible, although discrimination of sleep from wakefulness may be required for an accuarte calculation of the apnea/hypopnea index. Multichannel system allow detection of patients with sleep-disordered breathing, can accurately detect decreases in oxygen saturation, however some difficulty in classifying respiratory events has been reported (Gyulay S et al., 1986; Redline S et al., 1991). There has been more experience with multichannel recordings in adults than in infants and young children. Complex recording systems in children were used for the investigation of cyanotic or apneic episodes and UAW obstruction during sleep, but turned out to be more appropriate for use in a hospital settings (Abraham NG et al., 1990).

Actigraphy may be the subject of monitoring in a home setting. It was developed as a practical measure for long-term quantification of sleep/wake and circadian periodicity of movements (Sadeh A et al., 1991). A small motion sensor and digital recorder is usually worn on a wrist and later downloaded into a computerized analysis system. Accuracy for sleep-wake discrimination is 85 % to 93 % (Sadeh A et al., 1991). It is likely that a simple motion sensor combined with respiratory sensor and pulse oximeter may produce a diagnostic system for evaluation of OSAS in infants and young children.

Pulse oximetry is another tool of monitoring. However, it should be noticed that pulse oximetry alone is not sufficient to exclude the diagnosis of OSAS, because obstructive events in some patients may lead to arousals and sleep disturbance without significant desaturation. The method was characterized by high specificity but low sensitivity in diagnosis of OSAS (Ryan PJ et al., 1995). The method is good as a tool of home monitoring in infants and young children with BPD, cystic fibrosis and other chronic pulmonary diseases, as well as in patients with neuro-muscular diseases and myelomeningocele.

In some complicated cases, methods of *audio and video recordings* help to identify snoring and noisy breathing. Although this method does not allow to reliably distinguish between CA and OA (Postic WP, 1987), in combination with pulse oximetry it appears to be more effective, and may preclude the need for further PSG-study.

There are groups of infants for whom use of a home cardiorespiratory monitor may be most warranted. Infants for whom these indications may apply include (American Academy of Pediatrics, 2003):

(1) infants who have experienced an ALTE;
(2) infants with tracheostomies or anatomic abnormalities that make them vulnerable to airway comprimise;
(3) infants with neurologic or methabolic disorders affecting respiratory control; and
(4) infants with chronic lung diseases, especially those requiring supplemental oxygen, CPAP, or mechanical ventilation.

Table 47. Acid suppresants used in treatment of GER/GERD in infants (Arguin AL, Swartz MK, 2004)

Type of Medication	**Recommended Oral Dose**	**Adverse Reactions**
Histamine H-2 antagonists		
Ranitidine Syrup 15 mg/ml	Premature and term infants < 2 weeks: 2 mg/kg/day PO divided BID Infants > 1 month: 4 - 10 mg/kg/day PO divided BID Maximum dose for GERD: 300 mg/QD Maximum dose for erosive esophagitis: 600 mg/QD Administer with food Time to peak serum concentration: 1-3 hours	Fatigue, irritability, constipation, diarrhea, thrombocytopenia, bradycardia, tachycardia, sedation, headache, rash, elevated serum creatinine. Use with caution in infants with renal or liver impairment.
Famotidine Liquid 40 mg/5 ml	Infants: 1 mg/kg/day PO divided BID Maximum dose 80 mg/QD May administer with food Shake suspension vigorously before giving Maximum effect: 1-4 hours	Headache, dizziness, constipation, diarrhea, nausea, vomiting, bradycardia, tachycardia, thrombocytopenia, rash, elevated liver enzymes, BUN, and serum creatinine. Use with caution in infants with renal or liver impairment
Lansoprazole Enteric coated capsules: 15 mg, 30 mg; Granule packets: 15 mg, 30 mg	<10 kg: 7.5 mg once daily ≤ 10-30 kg: 15 mg once daily >30 kg: 30 mg once daily (may increase up to 30 mg BID if symptoms persist after 2 weeks of treatment) Reflux esophagitis: 30-60 mg once daily for up to 8 weeks Best if taken 30 minutes before meals Capsules may be opened and sprinkled in slightly acidic juice (e.g., apple, orange), applesauce, or yogurt. Contents should not be chewed or crushed. Packets of granules should be mixed only with water. Time to peak serum level: 1-2 hours	Diarrhea, abdominal pain, nausea, headache, hyper/hypotension, fatigue, dizziness, headache, rash, elevated serum transaminases, proteinuria.
Omeprazole Oral susp: 2 mg/ml; Capsules: 10 mg, 20 mg, 40 mg	1 - 3 mg/kg/day in two divided doses. Give before meals Shake vigorously before giving. Store in refrigerator Capsules can be opened and mixed with acid juice such as apple juice, cranberry juice, or applesauce. Maximum action: 2 hours	Diarrhea, nausea, vomiting, abdominal pain, constipation, bradycardia, tachycardia, headaches, dizziness, rash, agranulocytosis, elevated liver enzymes, proteinuria

The parents of such infants should be counseled regarding the purpose of home monitoring and realistic expectations of what it can and cannot contribute to an infant's well-being. Physician should also establish a specific plan for periodic review and termination. Should monitoring beyond 43 weeks' postconceptional (postmenstrual) age be recommended, clear documentation of the reasons for continuing monitoring is necessary (American Academy of Pediatrics, 2003). Before home monitoring is recommended, the physician should consider whether treatable causes for the apneic episode are present.

Parents and other caregivers must be trained in observation techniques, operation of the monitor, and infant cardiopulmonary resuscitation. Medical and technical support staff should always be available for direct or telephone consultation. Psychosocial assistance and respite personnel should also be available. A care plan including periodic reassessment of historical, physical, developmental, and laboratory data are essential. Long-term follow-up of neurodevelopmental status is advised (American Academy of Pediatrics, 2003).

There is no conventional monitoring protocol for the infants with AP and AI, however some practical guidelines have been offered (Spitzer AR, Fox WW, 1986).

Part 4. Sudden Infant Death Syndrome

Chapter XIV

Definition and Epidemiology

Sudden infant death syndrome (SIDS) is known as the sudden death of an infant under one year of age that remains unexplained after a thorough case investigation, including performance of a complete autopsy, examination of the death scene, and review of the clinical history (Willinger M et al., 1991). By definition, SIDS is a diagnosis of exclusion, and thorough autopsy should be performed in cases of sudden deaths in infants to rule out other possible causes of deaths. There is no "gold standard" of postmortem studies needed for correct distinction between the cases of SIDS and other causes of infant deaths so far, however several attempts have been made to offer standardized autopsy protocol. An example is International Standardized Report of the Autopsy (ISAP) (Krous HF, 1995).

By the end of the 1990-ies, the incidence of SIDS in the majority industrial countries across the world varied from about 0.1 to1 case per 1000 live births (Figure 11).

Much attention is paid to *social and demographic risk factors* of SIDS *characteristics.* Cases of SIDS are associated with low parental educational level, and based on population attributable risk, near 12 % of SIDS cases were associated with this factor (Kelmanson IA, 1994). Similar findings were offered by other authors (Hoffman HJ et al., 1988; Hoffman HJ, Hillman LS, 1992). Maternal marital status is another factor that influences the risk of SIDS, and babies from incomplete families (single, divorced or widowed mother) were at high risk with population attributable risk equal to 23,4 % (Kelmanson IA, 1994). High risk of SIDS is also associated with young maternal age (Cameron MH, Williams AL, 1986; Hoffman HJ, Hillman LS, 1992). Generally, some form of "social strain" is a common finding in the families where cases of SIDS occurred (Spurlock CW et al., 1987). Attention was focused on certain features of infant's microenvironment known to influence his/her development. Using standardized questionnaire (Pediatric Review of Children's Environmental Support and Stimulation, PROCESS), it was found that in cases of SIDS, the mothers often ignored the cry of their infants, did not care for organized nap time keeping, often had uncomfortable feelings when caring for their babies, and seemed to spend less time helping their baby to learn new things. As well, the mothers of unexpected deaths seemed more often to be indifferent to the health visitors, indifferent to the infant's behavior and progress, more reluctant to communicate with the baby and to carry the baby.

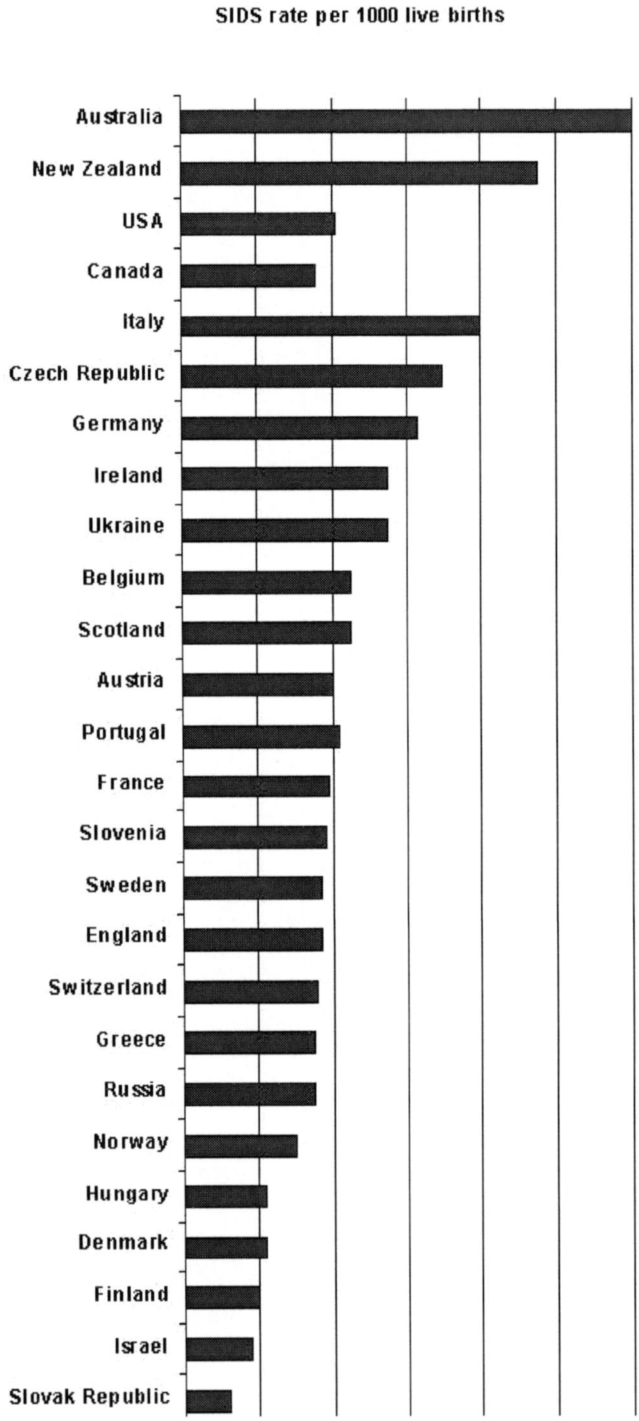

Figure 11. Mean for period 1997-2001 rate of SIDS in different countries (WHO data)

They were less eager to quiet or comfort the infant and paid less attention to the baby's responses while the pediatrician was examining the child. These mothers appeared to be often detached from the baby and have reduced eye-to-eye contact. They responded less to the infant's vocalization and often demonstrated negative attitudes to the baby. Such mothers were less responsive to praise of the child by the doctor and smiled and looked tenderly at the baby less frequently. The babies themselves were observed to be less clean. The summarized scores on developmental stumulation and environmental organization were lower in SIDS group (Kelmanson IA, 1993).

Table 48. Maternal smoking and the risk of SIDS (Kelmanson IA, 1994)

Smoking habit	Relative risk of SIDS (compared with non-smoking mothers, odds ratio)	95% confidence interval
Mother smoked 10 or more cigarettes daily		
Before pregnancy	0.96	0.56-1.64
During pregnancy	10.16	3.68-30.29
After delivery	2.42	1.28-4.61

One of the challenging issues is possible *association between maternal smoking and the risk of SIDS*. Table 48 is the summary of the own findings on that association (Kelmanson IA, 1994). Obviously, maternal smoking, particularly during pregnancy, is associated with increased risk of an infant to become a victim of SIDS, and according to the estimates of population attributable risk, 23.3 % of SIDS cases may be ascribed to the adverse effect(s) of maternal smoking during pregnancy (Kelmanson IA, 1994). Similar findings were obtained by the other authors (Table 49).

Risk of SIDS increases with increasing parity, short inter-pregnancy interval, high number of previous medical termination (Kelmanson IA, 1994).

Low maternal systolic BP during pregnancy ($\leq$ 109 mm. Hg) is associated with increased risk of SIDS (Naeye RL, 1983; Kelmanson IA, 1994). This low BP in combination with low maternal hemoglobin level may contribute to fetal hypoxia.

High risk of SIDS is associated with multiple births (Rintahaka PJ, Hirvonen J, 1986; Kelmanson IA, 1994). Short second period of delivery is also associated with increased risk of SIDS (Rintahaka PJ, Hirvonen J, 1986).

Recurrent infant mortality in siblings is another critical issue related to SIDS. The next siblings of first-born infants dying of SIDS are at increased risk for birth-related deaths (Oyen N et al., 1996). Similarly, the next sibling of first-born infants dying of birth-related deaths are at increased risk for SIDS (Kelmanson IA, 1994). When considering the specific cause of death in a subsequent sibling of an infant who died of SIDS, the relative risk for same cause of death was 9.1 compared with 1.6 for a different cause of death (Hunt CE, 2001).

The risk for recurrent SIDS in siblings is increased compared with the infants without previous history of SIDS in the families (OR from 3.7 to 10.0), however, is similar to increased risk for recurrence of non-SIDS causes of infant mortality (Guntheroth WG et al., 1990; Oyen N et al., 1996; Leach CEA et al., 1999). Although increasing parity is associated with higher risk for SIDS, the observed relative risk for recurrent SIDS is 4.6 times greater than expected by the effect of parity alone (Oyen N et al., 1996). The recurrence risk for all subsequent siblings is increased to a similar and significant extent as for the next sibling (Peterson DR et al., 1986; Oyen N et al., 1996). The realtive risk of SIDS was also similar when analyzed for prior siblings (Beal SM, Blundell HK, 1988).

Table 49. Maternal smoking during pregnancy and the relative risk of SIDS according to different sourses

Author(s)	Odds ratio (95% confidence interval)
(Bergman AB, Wiesner LA, 1976)	2,15 (1,02-4,53)
(Naeye RL et al., 1976a)	1,57 (1,02-2,42)
(Lewak N et al., 1979)	4,40 (2,10-9,86)
(Murphy J et al., 1982)	2,79 (1,79-4,36)
(Van den Berg M, 1985)	2,10 (1,49-2,96)
(Cameron MH, Williams AL, 1986)	3,90 (2,69-5,66)
(Rintahaka PJ, Hirvonen J, 1986)	4,12 (2,32-7,33)
(Hoffman HJ et al., 1988)	3,41 (2,81-4,12)
(Malloy MH et al., 1988)	2,92 (2,38-3,59)
(McLoughlin A, 1988)	3,29 (1,47-7,45)
(McGlashan ND, 1989)	1,50 (0,94-2,39)
(Gilbert RE et al., 1990)	2,44 (1,43-4,17)
(Haglund B, Cnattingius S, 1990)	2,35 (1,75-3,15)
(Wierenga H et al., 1990)	6,00 (1,20-38,57)
(Li DK, Daling JR, 1991)	2,98 (2,55-3,49)
(Mitchell EA et al., 1992)	4,09 (3,26-5,14)

Much attention should be paid to characteristics of the infants who died of SIDS. Boys are at higher risk of SIDS, and own data were that 59.4 % SIDS cases were males (Kelmanson IA, 1994). Similar findings were reported by other authors (Valdes-Dapena M, 1980).

More often than in general population future SIDS victims are born pre-term. Pre-term deliveries were observed in 29.7% of SIDS cases, and the relative risk of SIDS associated with prematurity was equal to 3.78 (95 % CI = 2,66 - 5,38). According to population atriburable risk, 18 % of SIDS cases were related to the effect of prematurity (Kelmanson IA, 1994).

Future SIDS victims are more often born with the signs of intrauterine growth retardation (IUGR). Signs of IUGR were found in 18.8 % of SIDS cases (Kelmanson IA, 1994). In high percent cases, these infants cannot catch up with their peers by the time of death, and have specific growth pattern of certain body segments (Kelmanson IA, 1991a; 1992). Remarkably, peak incidence of SIDS coincides with those periods of infant's growth that may be characterized as the periods of prevailing accretion of body weight and breast circumference compared with accretion of body lengths, i.e. coinsides with the periods of intensive cellular and tussie differentiation (Kelmanson IA, 1994).

Compared with healthy controls, infants who died of SIDS more often had lower Apgar scores at first and fifth minute, needed admission to NICU, required tube feeding. Such signs and symptoms as cyanosis and tachycardia in the neonatal period, increased body temperature, respiratory distress, irritability, episodes of hypothermia, tachypnea and apnea were more common for the future SIDS victims (Hoffman HJ et al., 1988; Hoffman HJ, Hillman LS, 1992).

Age period 2-4 months is known for peak incidence of SIDS. This is one of the specific epidemiological features of SIDS. Table 50 is the summary of own findings on age distribution of SIDS cases and other major causes of sudden, unexpected non-SIDS deaths in infants (Kelmanson IA, 1994).

SIDS cases have specific chronobiological features. The findings were that maxium risk of these deaths was between 04:00 and 06:00, meanwhile the lowest risk was during time interval 18:00 - 22:00 (Kelmanson IA, 1991b).

Certain infant-rearing practices may influence the risk of SIDS.

1. *Infant feeding.* Relative risk of SIDS in 2-4 month-old infants associated with bottle compared with breast feeding ranged from 1.89 to 3.30 (Kelmanson IA, 1994). The mechanisms underlying that association need to be further clarified. Attention was paid to toxigenic strains of E.coli that were found more often in the intestine of the bottle-fed infants (Bettelheim KA et al., 1990). Presumably, endotixins produced by these toxigenic strains may be absorbed, bind to CNS receptors of the critical sites responsible for respiratory control, and impaire respiratory function. At the same time, bottle-feeding is more common feature in the families with poor social and demographic status, and its seems more than likely that this is but a signs of inadequate infant microenvironmental characteristics.
2. *Heavy wrapping and overheating.* Several studies have shown that SIDS victims were more heavily wrapped than controls and were more likely to have had heating on through the night (Fleming PJ et al., 1990). Combination of a viral infection and heavy wrapping was associated with a very high relative risk of SIDS (Gilbert R et al., 1992).

   Thermal stress may lead to death by direct hyperthermia or act through central control mechanisms by impairement of respiratory drive, laryngeal closure reflex, depression of arousal mechanisms (Fleming PJ et al., 1993). There may be a developmental shift in vulnerability to thermal stress during early infancy, in that the newborn baby is particularly vulnerable to the effects of cold stress, but with advancing age heavy wrapping or a high environmental temperature become more critical. Disorder of thermoregulation may be aggravated, when the infan's head is covered during sleep. It was found that 18 % of SIDS victims compared with 2% of controls had their heads covered (Fleming PJ et al., 1996). Rectal and peripheral temperature rises in infants when the head is covered if there is a drop in metabolic rate (Marks K et al., 1985). Increased ambient temperature is associated with higher arousal threshold in the infants (Franco P et al., 2001). In experimental aminals, period of mild hyperthermia during or shortly after ischemia had similar effects on brain pathology to prolonged ischemia (Dietrich W et al., 1990). Proposed mathematical models showed that the other known risk factors of SIDS could potentiate reduced thermal discharge in the infants, and therefore thermal stress was considered as possible common pathway (Nelson EAS, 1995).

3. *Infant sleeping position.* The possibility that *the prone sleeping might be a contributory factor in cases of SIDS* was proved in the 1970-ies (Beal S, Blundell H, 1978). These findings were further corroborated by other researchers, and large number of case-control studies addressing the relationship between the risk of SIDS and infant sleeping position were published in the 1990-ies (Table 51). All of these studies showed a significant excess of deaths among infants sleeping in the prone position.

Table 50. Age at the time of sudden infant deaths in SIDS and non-SIDS cases (Kelmanson IA, 1994)

Cause of death	Number	Mean age, months	S.D.	S.E.M.	Quartiles for age distribution		
					Q1	Q2	Q3
SIDS	198	3.21	2.11	0.15	1.23	2.32	3.72
All non-SIDS cases	252	4.23***	2.94	0.19	1.38	3.18	5.34
Congenital anomalies and inherited disorders	64	4.54***	3.12	0.39	1.28	3.50	5.81
Castro-intestinal infections	16	3.63	3.01	0.75	1.0	2.0	3.99
Sepsis and neuro-infections	54	5.20***	3.29	0.45	1.89	3.99	7.87
Respiratory infections	118	3,75*	2,55	0,23	1,37	2,77	4,33

S.D.=standard deviation; S.E.M.=standard error of the mean.
Significant differences with mean age in SIDS group, t-test: * - $p < 0.05$; *** - $p < 0.001$

Table 51. Association between infant prone sleep and risk of SIDS

Author	Year	Region	Relative risk (odds ratio)	95% CI
(Fleming PJ et al., 1990)	1990	England	8.8	7.0-11.0
(Mitchell EA et al., 1992)	1992	New Zealand	3.7	2.91-4.70
(Ponsonby AL et al., 1992)	1992	Tasmania	4.58	1.48-14.11
(Fleming PJ et al., 1996)	1996	England	9.58	4.86-18.87
(Brook H et al., 1997)	1996	Scotland	6.96	1.51-31.97

Own findings were that the relative risk of SIDS associated with prone sleeping was equal to 2.25 (95 % CI = 1.01 – 5.01), which is in accord with the above results (Kelmanson IA, 1994).

These findings stimulated many public health services in most industrial countries to launch the campains aimed at propagation of putting babies to sleep on their back (the Netherlands -1987, New Zealand -1991, Australia -1991, Great Britain- 1991, USA - 1992). In the English-speaking countries, these actions have been conducted under the slogan "Back to sleep". The results were impressive and encouraging. For example, in the Netherlands, SIDS rate has been reduced from 1.04 per 1000 live-born infants in 1987 to 0.44 per 1000 live-born infants in 1991, and this reduction was not accompanied by major changes in the prevalences of another known risk factors of SIDS (de Jonge GA et al., 1993). Dramitic reduction of SIDS rate in many industrial countries that was achieved in the second half of the 1990-ies was in many respects due to propagation of public knowledge on potential benefits of sleep on the back. The practice of putting babies prone to sleep in the USA has been reduced from as high as 70 % in 1992 to 20 % in 2000, and, in parallel, the incidence of SIDS has been reduced by 40 % (American Academy of Pediatrics, 2000). Similar results were found by Norwegian authors (Irgens LM, 1995).

Local practice to put babies prone to sleep was the subject to analysis (Kelmanson IA, 1999a). Since local practice to place babies prone to sleep was rather uncommon, no wide intervention campaign against pronation has been launched, and the parents were likely to follow their and (or) infantile natural inclination when choosing baby's sleeping position. It was found that only 4 % of infants aged 2 to 4 months (peak incidence of SIDS cases) were routinely put to sleep prone (Table 52), and this was much lower than the reported prevalence figures on infant's prone sleep in many industrial countries.

Of 176 cases, 101 (57.4%) babies were usually put to sleep supine, 68 (38.6%) on the side, and 7 (4.0%) prone. The infants from three groups did not differ in sex distribution, birth weight, gestational age, age and weight at study. In 127 (72.2%) observations, infants were usually found in the same position as have been put to sleep; however in the remainder 49 (27.8%) they habitually changed position. Of 101 infants who were usually put to sleep supine, 93 (92.2%) were usually found on the back, 6 (5.9%) on the side, and 2 (1.9%) on the stomach. Of 68 babies who were usually put to sleep on the side, 27 (39.7%) retained that position, while 37 (54.4%) were usually found on the back, and 4 (5.9%) were usually found prone. Overall, of 169 babies who were usually put to sleep non-prone, 6 (3.6%) habitually turned on the stomach. Compared with infants put back to sleep, babies put side bore the relative risk to be found prone equal to 2.97 (95% CI: 0.56 - 15.77). None of the infants usually put prone to sleep modified that position. There was no statistically significant difference in any major infant and bedding characteristics between the babies who usually changed their initial sleep position and in those who did not.

Table 52. Relationship between sleep positions the infants are usually put to sleep and found awake, number observations (%) (Kelmanson IA, 1999a)

Found awake	Put to sleep			
	Prone	**Supine**	**Side**	**Total**
Prone	7 (4,0)	2 (1,2)	4 (2,3)	13 (7,5)
Supine	0 (0)	93 (52,8)	37 (21,0)	130 (73,8)
Side	0 (0)	6 (3,4)	27 (15,3)	33 (18,7)
Total	7 (4,0)	101 (57,4)	68 (38,6)	176 (100)

The propensity of a baby to modify his/her position during sleep and the pattern of that modification were found to be linked with such a trait of infant's temperament as persistence, and the lowest persistence was a feature of those babies who were usually put to sleep supine and found prone, while the highest persistence was characteristics of the babies put to sleep and found supine, and the differences in persistence scores between these two categories were most prominent (Kelmanson IA, 1999a). These findings could not have been ascribed to the exceptional influence of such potential confounders as gender, infant's age and maturity, as well as bedding, since there was no significant difference in any of the above characteristics between the infants who usually changed and did not change their position during sleep. During sleep, the infants are likely to spontaneously modify their position, and facing potentially unfavorable prone position, they probably can maintain adequate position and, respective, respiratory and metabolic control by turning head aside, vigorous body movements, spontaneous arousals, and other physiological mechanisms. Presumably, in some infants these protective mechanisms may fail due to neurological deficiencies, thus making them more likely to stay lying prone, should they happened to turn down in sleep. Perhaps, these neurological dysfunctions may also manifest as minor behavioral deviations presenting as low persistence. Another explanation may be that preferred sleep position can on its own somehow influence infant's behavior, particularly related to motor activity, which has been previously reported (Jantz JW et al., 1997). Moreover, repeated sleep apneas which may be more frequent during prone sleep, can jeopardize the integrity of the brain by short but frequent episodes of reduced oxygenation leading to impaired behavioral development (Einspieler C, 1994). However, under this assumption, it seems difficult to explain, why the babies usually put to sleep and found prone were rated as more persistent compared with those put back to sleep but found prone. Probably, all offered mechanisms may be involved and interact. Whatever the explanation, one can conclude that the probability with which infant's sleep position may be spontaneously modified during sleep is related to behavioral peculiarities, and that 2-4-month-old babies presenting as low persistence should be considered cautiously for increased risk to be found prone while put to sleep supine.

Spontaneous modification of sleep position during sleep was studied in healthy infants by the means of overnight PSG-recording and simultaneous videorecording (Kelmanson IA, 1997). The position "prone, face down" was registered in 4 of 9 studied babies. In 3 infants, this position was recorded only once during nighttime, and these episodes lasted 1.0, 3.3, and 6.5 minutes respectively. In one infant, two episodes have been recorded and lasted 0.1 and 0.2 minutes respectively. When these episodes occurred, infants were able to quickly turn head aside. All the episodes of prone, face down positions occurred during active sleep. Total time the infants spent in that position ranged from 0 to 6.5 minutes per night (mean value 1 minute), which made up from 0 to 5% (mean 1%) of the total time the infant was sleeping in prone position. The rest of the time the infants layed prone in sleep their face was turned aside (Table 53). The findings were that it was rather uncommon for the healthy infants to sleep prone and face down, and if these episodes occurred during active sleep, infants were able to quickly escape it by turning face aside.

The question as to how infant's prone sleep may influence the risk of SIDS still needs to be further clarified. In the prone position the infant loses less heat than in the supine position, partly as a result of greater flexion of the legs on the body and partly because the face, a major site of heat loss, is potentially in contact with the bedding. In the prone position, infants are slightly warmer and body temperature rises more rapidly during the latter part of the night than in the supine position (Tuffnell C et al., 1995). In prone position babies are also more vulnerable to the effects of rebreathing expired gases, particularly if sleeping on soft bedding (Bolton DP et al., 1993). As mentioned above, arousal threshold is higher in prone position (Franco P et al., 1998a), and hypoxia may have more profound and dangerous effects in infants sleeping prone. It was also suggested that in the infants lying prone cerebral blood flow may be compromised due to neck extension (Saternus KS et al., 1986). It was shown that high cervical stress can result in such autonomic reactions as hyperemia, apnea, hyperextension, and tachypnea (Koch LE et al., 1998). As well, mechanical irritation of the high cervical spinal cord region in infants may lead to cardiac arrythmia, including deep bradycardia (heart rate may drop as low as by 50-80% of the initial values); apnea and bradycardia may be often associated (Koch LE et al., 2002).

Table 53. Infant prone position at night according to overnight videorecording (Kelmanson IA, 1997)

Characteristics	Duration, minutes. Mean value (range)	% of the time
Total time the infant layed prone during sleep	211 (13-369)	100
Face turned to the right	121 (3-368)	49 (15-100)
Face turned to the left	89 (0-178)	51 (0-85)
Face down	1 (0-6,5)	1 (0-5)
Total time the infant layed prone while was awake	42 (0-113)	100
Face turned to the right	25 (0-87)	40 (0-95)
Face turned to the left	18 (0-51)	60 (6-100)
Face down	0,04 (0-0,3)	0,2 (0-1)

Another opportunity may be that prone, face-down position can predispose to certain changes in UAW configuration that in turn would predispose to disordered breathing. To test this hypothesis, twenty two infants that have been hospitalized for apneic episodes and who were ascertained to the cases of idiopathic ALTE after through examination were undergone the series of MRI in prone and supine positions (Kelmanson IA, 1997). The only significant differences found within configuration of the UAW in prone position compared with supine were the larger size of the tongue (+6 %), higher frequency of tongue protrusion, and more frequent observations on the contact between the tongue and soft palate. However, infants sleeping prone and face down on soft bedding show episodes of airway obstruction (Tonkin S, 1975).

It was also shown that in the prone position infants moved less, particularly during active sleep, than in the supine position and had higher heart rates and periferal temperature (Skadberg BT, Markestad T, 1997).

Another suggestion was a possible mechanism by which the prone sleeping position, heavy wrapping, and the presence of a viral infection might predispose to development of a secondary bacterial infection, with release of inflammatory mediators into the pharynx, leading to the rapid and potentially lethal development of shock (Blackwell CC et al., 1995).

4. *Infant-parent(s) bed sharing ("cosleeping").* Potential advantages and disadvantades of cosleeping is a controversial issue. Some authors have pointed out that the infant should be in close and more or less continuous contact with the primary caregiver, usually the mother (McKenna JJ et al., 1993). Studies of

communities in which cosleeping is the traditional approach have shown that when the infant is put down to sleep it is almost invariably in the supine position, and in such populations, the mother usually sleeps very close to the baby, very often in the same bed, and the incidence of SIDS is low (Davies DP, 1985; McKenna JJ et al., 1993). Some authors supposed that bed sharing might be potentially protective against SIDS, in part due to higher frequency and prolonged duration of breast feeding (McKenna J et al., 1994; McKenna JJ et al., 1997), increased environmental concentration of CO_2 during sleep with stimulating effect on infant's respiratory drive (McKenna JJ et al., 1997), changes in sleep structure with higher arousal frequency (McKenna JJ et al., 1997), as well as due to larger number of maternal inspections of an infant (McKenna J et al., 1994). Meanwhile, epidemiological studies showed that infant-parent(s) bed sharing was associated with increased risk of SIDS, particularly in the cases when the parents were smokers (Mitchell EA et al., 1992; Fleming PJ et al., 1996). Own findings were that, compared with the other sleeping routines, *the risk of SIDS was maximal, when the infant was put down to sleep in one bed with the parents, irrespective of parental smoking habits* (Table 54). The same conclusions were drawn by other researchers (Hauck FR et al., 2003), and according to recommendations of the experts of the European Society for the Study and Prevention of Infant Death, co-sleeping should be discouraged (Rognum TO, 2003). Own findings were that, from the standpoint of reduction of the risk of SIDS *the optimal practice should be to put an infant down to sleep in his/her own bed, but in the same room with the parents,*. Beneficial effect of room (rather than bed) sharing was mentioned by other authors (Scragg RKR et al., 1996).

5. *Use of a pacifier.* Another challenging issue is potential effect of the use of a pacifier on the risk of SIDS. There are findings that *the risk of SIDS may be reduced when the baby is put to sleep with a dummy:* odds ratio was equal to 0.3 (95 % CI = 0.2 – 0.5) (Hauck FR et al., 2003). These findings and potential underlying mechanisms are still debated. It was suggested that the dummy may prevent an infant to turn down to prone position during sleep (Fleming PJ et al., 1996). However, own findings were that the use of a pacifier in 2-4-month-old infants did not influence parental choice how put down the baby to sleep, neither influenced a propensity of a baby to spontaneously modify sleep position (Kelmanson IA, 2000c). Attention was also paid to so-called non-nutritive sucking associated with dummy use, and its was suggested that non-nutritive sucking could promote production of certain biologically active substances influencing sleep structure and infant behavior. Use of a pacifier was shown to decrease arousal threshold (Franco P et al., 2000a). Infants who used the dummies were characterized as having less episodes of unmotivated cry, less anxiety during day time and at night, and having less episodes of so-called colics (Neely CA, 1979). Own findings were that the infants who used dummies had more rhythmic behavior (Kelmanson IA, 1999b).

Table 54. Association between infant sleeping routine and risk of SIDS (Kelmanson IA, 1993; 1994)

Sleep routine	Relative risk of SIDS (odds ratio)	95% CI
In one bed with his/her sibling	3.48	0.23-102.98
In one bed with the parent(s)	8.70	1.1-209.36
In his/her own bed, in the separate room	1.77	0.26-12.36
In his/her own bed, in the same room with the parent(s)	1.0	-

Another important issue is *possible relation between SIDS and ALTE.* Although there is some similiarities in epidemiological features of these two conditions (Kahn A et al., 1995), *it is not proven that infants with an ALTE are really aborted SIDS cases.* The heterogeneity of both ALTE and SIDS infants renders comparison between both conditions difficult. Factors responsible for difficulties in comparing studies include varied types of presenting spells, varied terminology, inconsistent efforts to identify causes of the episodes, dependence on untrained observers, validity of parents to correctly perceive a true life-threatening event, and lack of follow-up programs. Some history-based studies failed to outline significant differences between both groups regarding preceding family or personal history, although ALTE were characterized by a 1 to 3 weeks younger age, and more favorable circumstances at the time of the event, including sleeping supine at the time of the event (Kahn A et al., 1984; Wennergren G et al., 1987; Mitchell EA, Thompson JMD, 2001). A relation between both conditions could concern a fraction of all SIDS cases, as up to 10% of future SIDS victims were found cyanotic or pale during sleep, some weeks before death (Mitchell EA, Thompson JMD, 2001), and that infants were reported who eventually died from SIDS after having presented an ALTE (Kelly DH et al., 1986). Own data were that previous episodes consistant with criteria of idiopathic ALTE were found in 4 % infants who died with diagnosis of SIDS (Kelmanson IA, 1994).

Chapter XV

SIDS and Sleep Disordered Breathing

Many epidemiological features of SIDS are the arguments that potential respiratory mechanism(s) of this syndrome are highly probable, and one of the initial suggestions about possible relation between SIDS and impaired respiratory control was made in the 1970-ies (Steinschneider A, 1972). According to own findings, in 35.7% of 50 infants who underwent overnight PSG-study and later eventually died of SIDS, total duration of obstructive apneas during active sleep exceeded 90-th percentile of the reference values, and in 40,0 % infants total duration of mixed apneas registered during acive sleep exceeded 90-th percentile of the referent values. In 44,4 % of observation, the number of obstructive apneas per 1 hour of active sleep exceeded 90-th percentile of the reference values (Kelmanson IA, 1994). Table 55 is the summary of the major differences in PSG-findings between the infants who later died of SIDS and control infants. Central and mixed apneas were more common in SIDS victims, and these babies had higher mean length of the central apneas during active sleep. These findings were generally in accord with the results of other authors (Kelly DH et al., 1986; Ward SLD et al., 1986), although in some prospective studies it was shown that future SIDS victims had fewer short central apneas than control infants, but more frequent mixed and obstructive sleep apneas (Kahn A et al., 1992a). Relative risk of SIDS associated with parental reported apenas in infants was found to be 1.86 (95 % CI = 1.12 – 3.09), and according to population attributive risk, 8 % of the SIDS cases could be associated with this clinical symptom (Mitchell EA, Thompson JMD, 2001).

Postmortem findings support the development of frequent hypoxic events, possibly related to airway obstructions. These include a thickening of the basement membrane of the vocal cords (Shatz A et al., 1991), increased size and mass of the tongue (Siebert JR, Haas JE, 1988), the presence of intrathoracic petechia in about 80% of the victims, excessive retention of periadrenal brown fat, increased extramedullary hematopoiesis, or abnormal astroglial proliferation, especially in the brain stem (Valdes-Dapena M, 1986). Additional postmortem evidence of premortem hypoxia included growth retardation, elevated blood cortisol levels, brain lactate and H+ concentrations, elevated hypoxanthine levels in the vitreous humor, pulmonary neuro-endocrine cell hyperplasia, or vascular endothelial growth factor in the cerebrospinal fluid (Rognum TO et al., 1988; Hunt CE, 2001). Specific morphological findings in the lungs of the SIDS victims have been reported, including reduced number of terminal bronchioli and their density (Beech DJ et al., 2001), increased thickness of alveolar-capillary barrier (Bajanowski T, Brinkmann B, 2000), retarded formation of elastine fibres (Butcovan D et al., 1997).

There were reports on right ventricular myocardial hypertrophy in the infants who died of SIDS (Naeye RL, 1983; Valdes-Dapena M, 1986), which might result from repeated and prolonged episodes of hypoxia. As mentined above, these findings are common in the patients with confirmed OSAS. Own findings were that the thickness of right ventricular myocardium was higher in the infants who died of SIDS comapared with matched healthy babies who died of accidents (Kelmanson IA, 1994) (Table 56). This was noteworthy, since the total weight of the heart was generally lower in the SIDS victims than in the control ones. Further studies also confirmed that, after adjustment for major potential modifying factors, the weight of the hearts in the SIDS vistims was on average by 1.6 gm (95% CI = 0.3 – 2.9 gm) lower than in the control babies (Kelmanson IA, 1996b).

Table 55. Major polysomnographic differences between SIDS victims and controls (Kelmanson IA, 1994)

Parameter	SIDS			Controls			P (t-test)
	n	Mean	S.D.	n	Mean	S.D.	
Total number of central apneas	39	65.0	88.6	100	43.5	35.4	0.04
Central apneas per 1 hour of sleep time	39	9.5	12.0	100	5.9	4.8	0.01
Mean central apnea length during active sleep, seconds	19	6.0	0.7	100	5.7	0.4	0.006
Total number of mixed apneas	50	1.8	2.7	100	1.0	1.8	0.01
Number of mixed apneas per 1 hour of sleep time	39	0.3	0.4	100	0.1	0.3	0.03

Table 56. Morphometric characteristics of the hearts in the infants who died of SIDS and in matched controls who died of accidents (Kelmanson IA, 1994)

Characteristics	Group	Mean value	S.D.	S.E.M.
Right ventricular myocardium thickness, cm	SIDS	0.26*	0.151	0.044
	Control	0.15	0.068	0.020
Left ventricular myocardial thickness, cm	SIDS	0.51	0.235	0.068
	Control	0.53	0.192	0.056
Weight of the heart, gm	SIDS	24.3	5.30	1.53
	Control	28.0	9.05	2.60

t-test for significance of the differences between the means in two groups: * $p < 0.05$

Infants who became victims of SIDS have been reported to arouse less from sleep than control infants. Cortical arousals were significantly less frequent in the future SIDS victims than in the control infants during both REM and NREM sleep. The frequency and duration of subcortical activations were significantly greater in the SIDS than in the control infants during REM sleep. Compared to the control infants, the SIDS victims had more frequent subcortical activation in the first part of the night, between 21:00 and midnight, and fewer cortical arousals during the latter part of the night, between 03:00 and 06:00. These data were suggestive of incomplete arousal processes in infants who eventually die at a time they were presumed to have been asleep (Kato I et al., 2003).

The reduced frequencies of cortical arousal in the future SIDS victims could result from structural or functional changes within the infants' arousal systems. Pathological changes described in SIDS victims include brain-stem gliosis (Naeye RL et al., 1989), hypoplasia (Matturri L et al., 2002) or apoptosis (Waters KA et al., 1999). The probability of an infant to have morphological signs of excessive brain stem gliosis decreased with infant's drowth and maturation, however glial proliferation was more protracted in the victims of SIDS (Kelmanson IA, 1994/95). Brain development and maturation is intimately associated with glial proliferation, since glial cells play important role in myelinogenesis (Mickel HS, Gilles FH, 1970), and the period of the first two years of infant's life is known for increased activity of glial cells (Herschkowitz N, 1989). It seems logical to assume that the common observations of glial cells proliferation may reflect retarded development and maturation of the brain structures in the infants who died of SIDS (Oehmichen M, 1990). At least in part, this may be due to brain hypoperfusion and chronic hypoxia, and similar morphological findings were reported in patients who died from congenital heart diseases with proven chronic hypoxia (Naeye RL, 1983). It was also found that the degree of brain stem gliosis in the victims of SIDS was associated with the fact of maternal smoking during pregnancy and the number of daily smoked cigarettes (Storm H et al., 1999).

There are also other signes of delayed neuronal growth and differentiation in the SIDS victims, including higher density of dendritic spines in the brainstem (Quattrochi JJ et al., 1985). These structures are most developed at approximately 34-36 wk of gestation. Later on, the spines become reduced in number, and this is presumably associated with maturation of respiratory control and settling of a sleep-wake cycle (Oehmichen M, 1990). Another findings were reduced number of cerebellar cortex cells in the victims of SIDS (Oehmichen M et al., 1989).

Functional changes could involve specific synaptogenesis or synaptic activities within various arousal systems such as the noradrenergic, serotonergic, dopaminergic, cholinergic, histaminergic or orexin binding sites (Peyron C et al., 1998; Kinney HC et al., 2001; Sawaguchi T et al., 2004). Abnormalities of serotonergic neurons were shown in the ventral medulla of SIDS victims, in brainstem structures associated with respiratory, cardiovascular and arousal controls (Kinney HC et al., 2001; Nattie E, Kinney H, 2002). Prenatal nicotine exposure reduces arousal response to hypoxia or auditory stimulation (Franco P et al., 1999). Such changes could favor maturational or permanent dysfunction of the infant's arousal process.

As discussed above, in healthy subjects, arousability increases across the night as a function of accumulated sleep time. It remains unclear why such a progressive increase in arousal frequency occurred in the control infants, but not in the SIDS infants. The increased number of arousals could be associated with endogenous characteristics of the future SIDS victims. It might also result from a sleep deprivation in the early part of the night. Compared to control infants, future SIDS victims had more frequent subcortical activation in the first part of the night. This finding is reminiscent of the report of more frequent fluctuations in breathing and heart rates during the early part of the night reported in future SIDS victims (Schechtman VL et al., 1992a). It is not known whether this increase in early-night subcortical activation resulted in sleep fragmentation, with a decreased tendency to arouse from sleep in the latter part of the night. Indirect evidence in adults associates autonomic changes with sleep fragmentation, and it may thus be hypothesized that subcortical activation induced sleep discontinuity and a decreased tendency to arouse from sleep by the end of the night.

Attention was paid to characteristics of the sighs in the future SIDS victims and control babies. Sighs were found in all sleep states, but were more frequently found in NREM sleep. The distribution of the sighs in the various sleep stages was not different between the two groups. Pre-apneic sighs were significantly more common in the future SIDS victims. No significant difference was found between the two groups of infants for the frequency of the other types of sighs. Transient increases in EEG frequencies following the sighs were observed significantly more frequently in the future SIDS victims. Oxygen blood saturation preceding the sighs was lower in the SIDS victims (Kelmanson IA et al., 2003). These findings somewhat contradict previous reports on infants victims of SIDS who were shown to have less (Kahn A et al., 1988) or as frequent sighs than control infants (Kahn A et al., 1992a). The apparent discrepancy could result from the types of sighs analyzed and differences in patient selection. Most other found characteristics of sighs are reminiscent of findings reported in infants and adults sleep studies. The observation of a lower blood oxygen saturation in the future SIDS victims, compared to the control infants, cannot be explained. It could, however, have contributed to the development of a greater frequency of sighs in the future SIDS victims. The apneas that followed the sighs could have been initiated by a rapid decline in carotid body chemoreceptor afferent discharge, due to the sigh-induced increase in arterial oxygen concentration and decrease in blood carbon dioxide. The association between sighs and apneas could however result from the complex interactions of factors, including peripheral chemoreceptor sensitivity, the degree of maturation and behavioral state, or central control mechanisms, including forebrain structures. As mentioned before, association between sighs and reticular formation-related arousal mechanisms has been suggested. The greater frequency of increases in EEG frequencies seen in the future SIDS victims following the sighs preceding a central apnea could reflect a sleep-maintaining process. This finding again could be reminiscent of the reduced arousability from sleep reported to occur in victims of SIDS.

Studies in siblings of prior SIDS victims are potentially useful in uderstanding its mechanism(s). SIDS siblings are significantly more likely to have central and obstructive apneas and periodic breathing during sleep (Kahn A et al., 1986a). Infants with a family history of multiple SIDS, ALTE, or obstructive sleep apnea syndrome were more likely to have mixed and obstructive apnea and these apneas occurred more frequently (McNamara F, Sullinvan CE, 2000). Meanwhile, other studies failed to corroborate these findings (Flores-Guevara R et al., 1986; Hoppenbrouwers T et al., 1993). SIDS siblings had a normal ventilatory response to 4% CO_2 but responded to mild hypoxia (17% O_2) with decreased respiratory rate, increased total duration of apneas lasting more than 3 seconds, increased periodic breathing, and increased time spent in apnea compared with control subjects (Brady JP, McCann EM, 1985). SIDS siblings have increased tendency to remain asleep or a relative failure to arouse from sleep (Harper RM et al., 1981). A study of temporal patterning of slow-wave sleep observed significantly increased delta amplitude in the early morning at 3-4 mo of age, consistent with higher arousal thresholds (Schechtman VL et al., 1995). SIDS siblings were significantly more likely to stay asleep longer and to proceed from active sleep to quiet sleep, and than to awake (Hoppenbrouwers T et al., 1989). The number of SIDS siblings arousing to 4% CO_2 was not decreased compared with control subjects, but there was a trend toward fewer arousals to 17% O_2 (Brady JP, McCann EM, 1985). Increased brain stem auditory conduction times are indicative of delayed or deficient maturation of neural processing in the brain stem, and brain stem conduction times were prolonged in SIDS siblings (Pettigrew AG, Rahilly PM, 1985). Beta-endorphin levels in spinal fluid were significantly higher in SIDS siblings (Myer EC et al., 1987). Skin temperature in SIDS siblings was significantly increased during sleep, suggesting differences in temperature regulation (Hoppenbrouwers T, Hodgman JE, 1992). Although there are no other convincing evidence that autonomic function is markedly different in SIDS siblings, it was found that heart rate was increased overall in SIDS siblings compared with control subjects during overnight PSG-recording (Harper RM et al., 1978a).

It has been previously discussed that respiratory disturbances are commonly associated with disturbances of cardiac rhythm, and cardiac mechanisms are involved in the pathogenesis of sleep-disordered breathing. Future SIDS victims exhibit symptoms during sleep that reflect a subtle dysautonomia. These include episodes profuse sweat productions during sleep (Kahn A et al., 1986b), tachycardia, bradycardia, higher overall heart rate, or a reduced heart rate variability (Schechtman VL et al., 1992b). Analysis of the heart rate variability in future SIDS victims showed findings compatible with a decrease in parasympathetic, an increase in sympathetic activity, or a combination of both conditions. Future victims of SIDS have a higher peak of sympathetic tonus, desynchronized from the parasympathetic peak activity during the late hours of the night, when most sudden deaths occur (Franco P et al., 1998b). Such imbalance in cardiac autonomic control has been postulated to induce prolongation of the QTc interval in SIDS victims (Schwartz PJ, Segantini A, 1988). The greater sympathovagal control seen in infants at risk of SIDS could result from a delayed maturational process, from repetitive hypoxia, that modify brainstem, cerebellar or cortical differences in autonomic controls (Harper RM, Bandler R, 1998).

Autonomic cardiac controls are also dependent on environmental factors. Increases in sympatovagal controls have been measured following the prenatal exposure to cigarette smoke (Franco P et al., 1999), prone sleep (Franco P et al., 1998a), previous sleep deprivation (Franco P et al., 2003), sleeping in high ambient temperatures (Franco P et al., 2000b), or with the face covered by a bed sheet (Franco P et al., 2002a). Breastfeeding and the use of a pacifier, two factors associated with a lower risk for SIDS, were characterized by a reduction of the heart rate sympathovagal ratio (Franco P et al., 2004).

Chapter XVI

Infant Behavioral Characteristics Associated with SIDS and Its Risk Factors

High probability of minor neurologic disorders associated with respiratory disorders in infants who died of SIDS makes a search for behavioral deviations in these babies and infants at risk highly probable. A series of own studies has been undertaken to address this issue.

1. SIDS Victims

Temperament characteristics were studied in 36 infants aged 2-4 months in whom full post mortem and clinical studies had been carried out leading to a diagnosis of SIDS (Kelmanson IA, 1996a). The families were approached contemporaneously, 2 to 4 weeks after infant's death, and the mothers were asked to complete Early Infancy Temperament Questionnaire (EITQ) (Medoff-Cooper B et al., 1993). The control group was taken from the practice of the same pediatricians who have cared for the SIDS babies. A control clinically healthy child matched for sex, age, geographical distribution, and the date of birth as close as possible was taken from each pediatrician's outpatient clinic. The request to them was for participation in a scientific study, and the mothers filled in the same forms as the SIDS mothers.

The features that presented were that the infants who died of SIDS have moved less when lying awake in crib, during feeding, and during the nail cutting; less often they turned head away and looked for mother when held by new person; more frequently these babies were reported to show extreme response (either almost never protested or almost always objected) when approached by someone other than main care giver; as well, extreme behavioural patterns were noticed looking at infant's vocalising on waking up. The infants who died of SIDS had lower activity scores.

Lower activity, abnormal reactions to the new persons and specific behavior after sleep in the SIDS victims suggest a relation to the function of central nervous system. This finding is also in accord with reported data where parents retrospectively rated their infants who had died of SIDS as being less physically active, more breathless and exhausted during feeding, and having less intense reactions to environmental stimuli and more abnormal cries than their siblings or peers from general population (Naeye RL et al., 1976b; Einspieler C et al., 1988).

2. Microenvironment

Much interest has been raised in potential modifying effect of environmental factors (one of the major factors influencing the risk of SIDS) and the quality of parental care on infant's behavior. While some authors found little evidence that early, extensive, and continuous care modifications were significantly related to problematic behaviors (The NICHD Early Child Care Research Network, 1998), the others found that the total number of life events was a significant predictor of changes in the level of cognitive and social-emotional functioning (Ihle W et al., 1992), and even short-term infant-parent separation could provoke dramatic changes in infant affect, motor patterns, and appetite (Armstrong KL et al., 1998). Associations between environmental characteristics and infant attention (DiLalla LF, Bishop EG, 1996), mental and psychomotor development (Grantham-McGregor SM et al., 1998) also have been reported.

Possible relationship between the objective measures of the quality of infant care and behavioural characteristics in 2-4-month-old infants were studied in 250 randomly selected, apparently healthy infants (Kelmanson IA et al., 2002b). Quality of infant care was estimated using the "PROCESS" inventory (Pediatric Review of Children's Environment Support and Stimulation) (Casey PH et al., 1988). Associations were found between developmental stimulation and such infant's behavioral features as mood and distractibility, in that the babies facing more developmental stimulation had more positive mood (i.e., exhibited more pleasant behavior during the day) and were more distractible (i.e., needed less external stimuli, such as sounds, toys, people, etc., in distracting or soothing the baby and interfering with ongoing behavior). Organization was also significantly associated with infant's mood and distractibility, in that more organized infants presented with more positive mood and more distractible behavior; as well, higher organization was associated with higher rhythmicity (more regular schedule for sleeping, bowel movements, feeding, etc.), higher persistence known as an ability of a baby to maintain specific activity, and higher adaptability (less time needed for a baby to adjust to new routine, new people, new circumstances). These associations remained significant after adjustment has been made for major potential confounders.

The findings were in accord with previous reports on significant relationships between child behavior ratings and the home environment (Koniak-Griffin D, Verzemnieks I, 1995). Infant's development unfolds in an ecologic and a social setting which, just like the genes, is decisive in shaping the organism, and behavior is fashioned by social experience (Eisenberg L, 1999) with sensory, nutritive, endocrine, and neural mechanisms involved (Rosenblum LA, Andrews MW, 1994; Florey CD et al., 1995; Fleming AS et al., 1999). Even minor deficit in the quality of infant care obviously may be associated with specific disadvantageous behavioral features in the infants, typical for what is known as "difficult temperament». These features may probably reflect certain impairments in brain maturation, and some studies offered arguments on age-dependent causal relationships between brain development and infant behavior (Herschkowitz N et al., 1997). Less likely seems a "reverse causality", in that the infants with problematic behavior (moody, more arrhythmic, less distractible, slower adaptive and less persistent) could affect parental attitude to such a degree that provoked less developmental stimulation and organization.

Behavioral features related to low developmental stimulation and organization were somewhat different from those reported in the infants who died of SIDS. Obviously, behavioral peculiarities of the SIDS victims cannot be exceptionally ascribed to the influence of improper environment, although some interaction with the latter may be certainly involved.

3. Maternal Smoking in Pregnancy

The long-term effects on the subsequent behavioral and cognitive development of the infant and child are less clearly understood. Although studies in animal models have revealed lasting behavioral and cognitive impairments in offspring prenatally exposed to nicotine (Ajarem JS, Ahmad M, 1998), these effects in humans have not been well established (Chiriboga CA, 1993; Lassen K, Oei TP, 1998). While some authors failed to prove that either psychomotor or mental development of an infant was affected by maternal smoke exposure during pregnancy and early infancy (Schulte-Hobein B et al., 1992), the others found certain associations between antenatal smoke exposure and particular behavioral features in infants, including less optimal Brazelton neonatal behavioral assessments (Oyemade UJ et al., 1994), abnormal, pitched neonatal infant cry (Nugent JK et al., 1996), colic-like behavior (Haggart M, Giblin MJ, 1988), hypertonicity and increased nervous system excitation at one month of age (Fried PA et al., 1987), impaired babbling ability at eight months (Obel C et al., 1998), lower mental scores and altered responses to auditory items at one year (Orlebeke JF et al., 1997). The question whether, to what extent and in which direction maternal smoking in pregnancy may influence infant's temperament has been addressed to a lesser extent so far.

The study has been undertaken aimed at assessing possible association between maternal smoking in pregnancy and temperament in 2-4-month-old infants (Kelmanson IA et al., 2002a). Of 250 mothers comprised by the study, 64 women (25.6%) smoked during pregnancy. The infants born to smoking mothers were characterised by certain behavioral features including more frequent fussy periods occurring at about the same time of the day, more frequent protesting behavior at face washing and washing in bath, more frequent indifferent attitude to the mother when held by new person, more frequent extreme reactions (either indifference or much feeling) during diapering and bowel movement, less attention to the parents during parent-infant play activity, and more sensitivity to the wet diaper. These particular behavioral characteristics found in the infants born to smoking mothers were somewhat reminiscent of the reported association between maternal smoking in pregnancy and more frequent colic-like behavior in the babies (Haggart M, Giblin MJ, 1988). The infants born to the smoking mothers were characterized by more intensive reactions (i.e., displayed more amount of energy regardless of positive or negative behavior) compared with the babies from the non-smoking group. Linear regression analysis found significant "dose-response" effect between the mean number of cigarettes daily smoked during pregnancy by the mother and infant's intensity score. These associations remained significant after adjustment has been made for major potential confounders and modifying factors. The findings did not conflict with previous reports on infants increased nervous system excitation associated with maternal smoking in pregnancy (Fried PA et al., 1987).

4. Low Birth Weight

Another well-known risk factor of SIDS is low birth weight (LBW). Infants born LBW may face combinations of various neurosensory, developmental, and health problems that compound the clinical and educational outcomes. The study has been undertaken aimed at assessment of possible effect(s) of LBW on temperament features in the infants under one year old (Kelmanson IA, 1999d). The features that presented were that the LBW infants more often rejected new baby sitter, changes in foods, and protested against disliked medicines; more frequently they showed extreme behavioral patterns (either were very quiet or very anxious) when falling asleep and waking up. The LBW infants were more withdrawing and less adaptable.

5. Infant-Parent(s) Bed Sharing

Little is known on the fact whether, to what extent and in what direction parent-infant bed sharing may influence infant integral behavioral characteristics. To address the issue, the study was undertaken aimed at evaluating possible association between parent-infant bed sharing and behavioral features of 2-4-month-old infants (Kelmanson IA, 1999c). Of 204 infants who entered the study, 175 (85.8%) were solitary sleepers, 17 (8.3%) usually slept with one and 12 (5.9%) with both parents. Solitary sleeping infants were characterized by more positive mood compared with those who shared the bed with the parent(s), and most negative mood was the feature of those infants who shared the bed with both parents. Solitary sleeping babies appeared to be more persistent than the infants who shared the bed with the parent(s). These associations between sleeping habit and infant's mood and persistence remained after adjustment has been made for major potential confounders. It seems most logical to suggest that bed sharing has real causal effect on infant's behavior. According to previous findings, regular co-sleeping with the parents led to more disruptive sleep in the infants and children, including difficulty in getting to sleep and increased night waking (Madansky D, Edelbrock C, 1990; Lozoff B et al., 1996), as well as enhanced arousals (McKenna JJ et al., 1997). On its turn, brief sleep was shown to be associated with negative mood in the infants (Weissbluth M, 1981; 1982a). Perhaps, similar speculations may be true for less persistent behavior associated with bed sharing. Another suggestion may be a reverse causality, in that some infant's behavioral features may stimulate the parents to put a baby to sleep with them. It seems logical to suppose that the parents may be keener to share the bed with an infant having more negative mood who demands more parental efforts to be soothed.

However, revealed "dose-response" effect, in that the infants who sleep with both parents presented as having more negative mood compared with the infants who shared the bed only with one parent, is an argument against reverse causality. Likewise, reverse causality cannot explain the finding on less persistent behavior in the infants who shared the bed with the parent(s). Whatever explanation, there were no any beneficial effects of parent-infant bed sharing on infant's behavioral features. In particular, bed sharing was not associated with increased infant's activity, which might have been considered protective against SIDS taking into account behavioral features of SIDS victims.

6. Use of a Pacifier

It has been reported that pacifier users had higher scores on each of the Bayley mental and motor scales at two years (Anderson GC, 1986). The study was performed aimed at the assessment of possible association between use of the pacifier and certain features of temperament in 2-4-month-old infants (Kelmanson IA, 1999b). Of 192 infants who made up a study group, 117 (60.9%) reportedly used a pacifier, while 75 did not. The findings were that, compared with non-users, pacifier users presented with higher rhythmicity, and this association remained significant after adjustment has been made for major potential confounders. Although revealed association between the use of a pacifier and rhythmicity was quite a new finding, it was in accord with the fact that infants who used a pacifier had earlier and more organized sucking patterns and more rapid gastrointestinal transit with more bowel movements (Anderson GC, 1986). Perhaps, a pacifier regularly offered to and withdrawn from the baby may serve to some extent as an external "organizer". This effect also can be modulated through the fluctuations in hormonal levels with increased secretion of gastrin and insulin during pacifier sucking. Another suggestion may be a reverse causality - more rhythmic infant may be keener to accept and/or demand a pacifier from the parents. It has been previously suggested that more passive and quiet infant who is not such a disturbance to the parents is less likely to be given a pacifier (Mitchell EA et al., 1993). On the other hand, infants who used a pacifier were reported to change less often their sleep position, thus appearing to be less active during sleep, and this behavioral feature was implicated for possible protective effect of a pacifier against SIDS due to reduced risk of a baby to turn in sleep to hazardous prone face-down position (Fleming PJ et al., 1996). However, own findings failed to confirm any association between the use of a pacifier and infant's activity.

7. Repetitive Regurgitation

Since pathological GER has been implicated for increased risk for SIDS (MacFadyen UM, 1993; Freed GE et al., 1994; Faubion WA, Jr., Zein NN, 1998), the question about possible association between infant's regurgitation and behavior becomes challenging. Behavioral deviations in the infants with repetitive regurgitation may pose additional problems for parents and/or care givers when handling the baby and affect mother-child interaction (Mathisen B et al., 1999). Certain behavioral features may also serve as early indicators of possible pathology. Some authors were uncertain whether even pathogenic GER may be likely cause of behavioral disturbances in early infancy (Heine RG et al., 1995), to say nothing of its relatively benign forms. The study was performed aimed at evaluation of possible association between maternal-reported repetitive regurgitation in otherwise healthy 2-4-month-old infants and their features of temperament (Kelmanson IA, 2000b). Repetitive regurgitation was statistically significantly associated with infant's distractibility: symptomatic infants were characterised as less distractible. This association remained significant after adjustment has been made for potential confounders. Such a behavioral feature that implied the need for more parental efforts to soothe the baby was in accord with reports on irritability, fretfulness, excessive crying, distressed behaviour in infants with GER (Gibson WS, Jr., Cochran W, 1994; Glassman M et al., 1995; Catto-Smith AG, 1998). It seems logical to suppose that repetitive regurgitation, even mildly expressed and unlinked with other major health problems, may affect infant's behavior. There are evidence for particular changes in the levels of neuroactive peptides such as beta-endorphines in some infants with confirmed GER (Beyaert C et al., 1995). Alternative opportunity may be a "reverse causality", in that fretful infant may have a higher frequency of regurgitation, particularly due to incompetence of lower oesophageal sphincter combined with the increased intraabdominal pressure from crying or straining (Fonkalsrud EW, Ament ME, 1996). Meanwhile, this latter suggestion cannot readily explain, why the infants with reported repetitive regurgitation did not differ from asymptomatic babies in other than distractibility "problem behavior" categories.

The data presented are highly indicative that the infants who died of SIDS were characterised by certain behavioral features related to temperament. Obviously, some of these features may involve age-dependent correlations of brain and behavioral maturation and involve causal relationships between brain development and behavior in the infants (Herschkowitz N et al., 1997). However, behavioral patterns associated with SIDS were found in a substantial proportion of control infants, and differences revealed between the infants from the case and control groups were not that striking. Therefore, behavioral symptoms could hardly be regarded as the reliable indicators of an increased risk of SIDS, although some of them (especially reflecting infant's activity when awake and his/her attitude to the mother) should be taken into account cautiously, and may serve as indicators for future clinical and physiological study. Certain endogenous and exogenous factors influencing risk of SIDS were shown to be associated with specific features of infant's behavior. While some of these features were highly reminiscent of what has been noticed in the SIDS babies, they were not exactly the same, and it seems logical to suppose that the traits of temperament characteristic for SIDS cannot be exceptionally ascribed to the modifying effect(s) of the considered risk factors. Perhaps, more complicated mechanisms are involved and interact which should be the subject of further investigations.

Certain features of infants temperament are associated with specificity of their autonomous reactions. In particular, infants with increased vagal tone present as less negative and less distractible by external stimuli, and the easiness with which increased vagal tone can be reduced is associated with increased span of attention periods and higher distractibility (Huffman LC et al., 1998). Characteristics of infant temperament may be linked with energy expenditures. Thus, the ability of an infant to gain excessive weight is more common in infants with negative mood (Carey WB, 1985), and infants characterized by high distractability have lower thickness of subcutaneous fat during further age periods and are characterized later by higher motor activity (Wells JC et al., 1997). Some biologically active substances can influence characteristics of infant's temperament. Plasma level of progesterone and its metabolites, which are known as CNS depressors, can influence infant's approach and intensity, and infants with "easy" temperament have higher plasma level of these substances than the infants characterized as "difficult" temperament (Weissbluth M, Green OC, 1984).

Attempts failed to reveal association between such characteristic of infant temperament as threshold (i.e. minimum level of external stimuli needed to avoke infant's reaction) and the threshold of arousal during a sleep, both in clinically healthy infants and in the infants with high risk of SIDS (Parslow PM et al., 2002).

There are data that so-called "difficult" temperament of an infant is associated with increased probability of household traumas, functional abdominal pains, repeated vomits, and episodes of so-called colic (Carey WB, 1972; Thomas A, Chess S, 1977; Carey WB, 1981; 1986; Super CM, Harkness S, 1986).

Fear parlysis (tonic immobility) have been implicated for possible mechanisms of SIDS (Kaada B, 1987). The response termed *"fear paralysis reflex"* consists of an immediate inhibition of all ongoing somatic movements, including a decrease in respiration (in expiration), a profound bradycardia, and an increase of peripheral vasoconstriction. This reflex may be provoked by such stimuli as restrain of movements, sudden exposure to unfamiliar environments and persons. In case when infant's movements were restrained, the exposure to noise of 100 dB during active sleep more readily provoked bradycardia (Kahn A et al., 1992c). Exposure to noises during sleep often causes irregular breathing sometimes degrading to apneas, and also may provoke bradycardia (Anderssen SH et al., 1993).

Chapter XVII

Model of SIDS and Preventive Strategies

It has been suggested that SIDS results from the intersection of three overlapping factors (*triple model*):

1. a vulnerable infant;
2. a critical developmental period in homeostatic control;
3. an exogenous stressor (Filiano JJ, Kinney HC, 1994).

This may be also described as *"three S"* (Figure 12):

1. *S*icknesses,
2. *S*tages of development, and
3. *S*urroundings.

An infant will die of SIDS only if he/she possesses all 3 factors, although the degree with which each specified constituent is expressed in a certain given baby, obviously may vary. The infant's vulnerability lies latent until he/she enters the crucial period and is subject to an exogenous stressor.

Infant's vulnerability includes risks that originate during pregnancy, but are not specific to SIDS. They are not controversial, because they consist of well-established epidemiologic risk factors for infant mortality in general, and they were described above. Four lines of evidence for prenatal origin of SIDS have been provided (Kinney HC et al., 1992):

1. Subtle CNS and/or systemic abnormalities in SIDS victims detected by quantitative or rigorous methods at autopsy;
2. Neonatal abnormalities in neurologic or autonomic function in at least some infants who subsequently die of SIDS;
3. Postneonatal abnormalities in cry, cardiac and ventilatory patterns, and state organization in some future SIDS victims, and
4. Maternal and pregnancy-related factors associated with increased SIDS

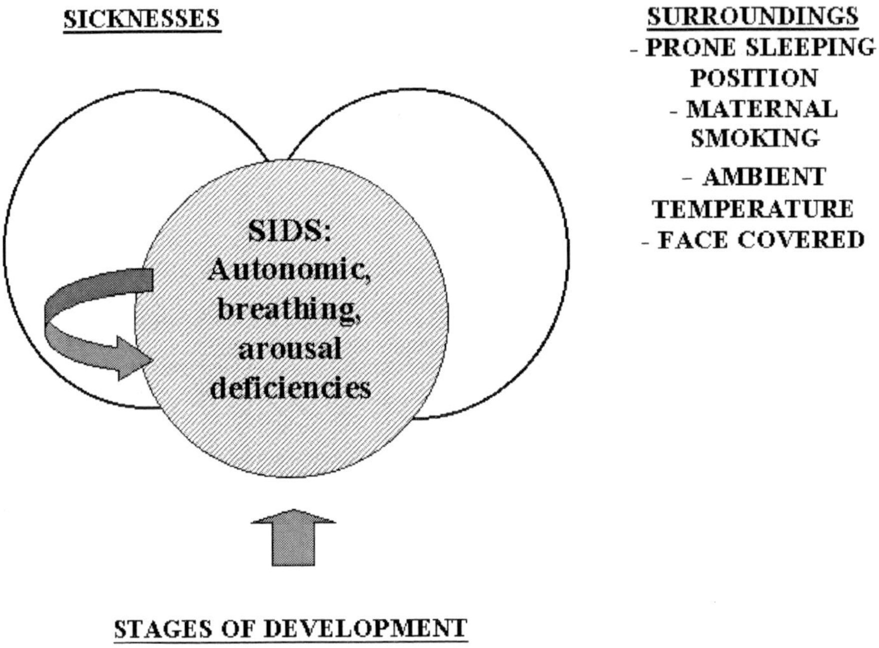

Figure 12. Model for SIDS

These risk factors point to a suboptimal intrauterine environment, and suggest that the mechanisms for risk for SIDS develop in fetal life in at least some cases (Kinney HC et al., 1992; Filiano JJ, Kinney HC, 1994). That fact that maternal mutritive status, hormonal level and environmental factors may cause long-lasting effects on the fetus and on the infant and child in postnatal life has been well established, and the term *"programming"* has been introduced to designate the process by which these adverse antenatal factors at the critical stage of the fetal development may lead to long-lasting or permanent consequences (Lucas A, 1991).

Infants who eventually die of SIDS may appear normal clinically, but their vulnerability lies latent until they enter the *critical developmental period*, when major changes occur in virtually all physiologic systems. The physiologic changes that occur from 1 to 4 months of life are incremental, and not nearly so severe and abrupt as the dramatic changes that occur at birth, and the first month is an uncommon time for SIDS. An alternative explanation of the age at death distribution for SIDS, particularly partial sparing of the first month of life, could be an early protective mechanism that is limited in duration. For example, gasping is capable of restoring regular respiration in neonates. The notable ability of the fetus for anaerobic metabolism is obviously still present in the neonate, but gradually diminishes in the first few weeks. However, this alternative explanation for the relative sparing of the very young infant from SIDS also postulates the possession of an innate but temporary protective mechanism that disappears with increasing age (Spiers PS, 2000). This mechanism is hypothesized to be negatively associated with the infant's level of maturity, thus accounting for the age of maximum risk for SIDS increasing as gestational age increases.

Exogenous stressors (precipitating events) are accepted by almost every hypothesis relating to SIDS. These precipitating factors, such as respiratory infections, have been identified by epidemiologic studies, and are not usually lethal because they are common occurrences in thousands of healthy infants who do not succumb to SIDS. Potential effect of major environmental factors known to influence the risk of SIDS is summarized on the scheme (Figure 13).

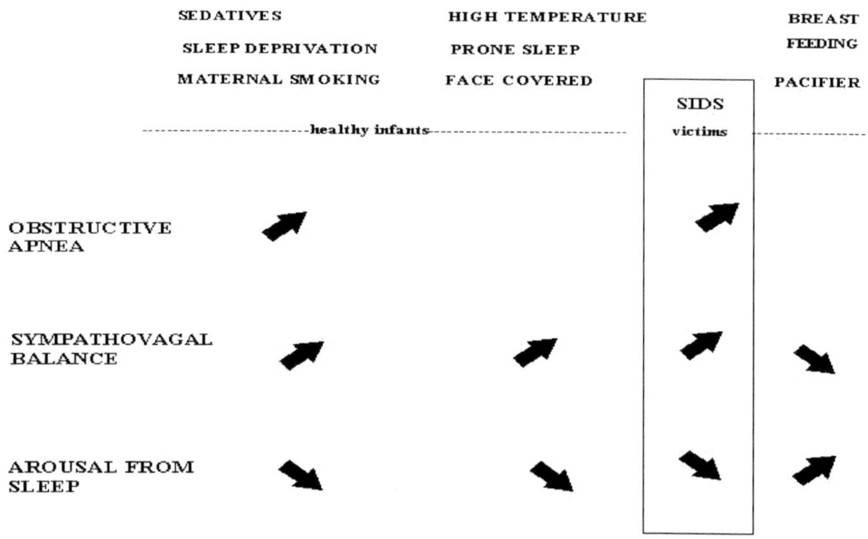

Figure 13. Infant and environmental stressors and their effects on vital control mechanisms

Most environmental risk factors of SIDS described in the previous section are modifiable risk factors. They are present in 30 to 80 % of future SIDS victims. Their avoidance contributes to the development of safe sleep environments and reduces the risk for SIDS by the continuous information of health professionals and the public. Most reasonable infant-rearing recommendations should include:

1. To avoid meternal smoking during pregnancy and smoking in the ambiance of an infant after birth;
2. To maintain breast feeding, at least during the first four months of infant's life;
3. To avoid heavy wrapping and overheating of an infant;
4. To put an infant to sleep supine, on a firm mattress, without excessive bedding;
5. To put an infant to sleep in a separate bed in the parental room (room sharing without bed sharing)

Special efforts must be addressed to populations at higher risk, such as those poorly influenced by usual prevention campaigns. These include the mothers of low education or families of lower socioeconomic status.

Epidemiologic studies have failed to document any impact of home cardiorespiratory monitoring for apnea and/or bradycardia on the incidence of SIDS, and prevention of SIDS is not an acceptable indication for home cardiorespiratory monitoring (American Academy of

Pediatrics, 2003). However, as previously discussed, cardiorespiratory monitoring may be prescribed for some infants with known risk factors for SIDS, including infants who have been born preterm, infants who have experienced ALTE, as well as to the siblings of the SIDS victims.

References

AASM. *International classification of sleep disorders, revised. American Academy of Sleep Medicine.* Westchester, IL 2000.

Abraham NG, Stebbens VA, Samuels MP, Southall DP. Investigation of cyanotic/apneic episodes and sleep-related upper airway obstruction by long-term non-invasive bedside recordings. *Pediatr. Pulmonol.* 1990;8:259-262.

Achermann P, Borbély AA. Simulation of human sleep: ultradian dynamics of electroencephalographic slow-wave activity. *J. Biol. Rhythms.* 1990;5:141-157.

Adair R, Bauchner H, Philip B et al. Night waking during infancy: the role of parental presence at bedtime. *Pediatrics.* 1991;87:500-503.

Adair RH, Bauchner H. Sleep problems in childhood. *Curr. Probl. Pediatr.* 1993;23:147-170.

Adam K, Oswald I. Protein synthesis, bodily renewal and the sleep-wake cycle. *Clin. Sci.* 1983;65:561-567.

Adulas EI, Kelmanson IA. An impact of massage therapy on sleep characteristics in infants born low birth weight. *Somnologie.* 2005;9(Suppl. 1):34.

Agarwal KN, Gupta A, Pushkarna R et al. Effects of massage and use of oil on growth, blood flow and sleep pattern in infants. *Indian J Med Res.* Dec 2000;112:212-217.

Ajarem JS, Ahmad M. Prenatal nicotine exposure modifies behavior of mice through early development. *Pharmacol Biochem Behav.* 1998;59(2):313-318.

Albani M, Bentele KH, Budde C, Schulte FJ. Infant sleep apnea profile: preterm vs. term infants. *Eur. J. Pediatr.* 1985;143(4):261-268.

Aldrich MS. Narcolepsy. *Neurology.* 1992;42 (Suppl. 6):34-43.

Aldrich MS. Diagnostic aspects of narcolepsy. *Neurology.* 1998;50 (2 Suppl. 1):S2-S7.

Ali MM, Russell N, Awada A, McLean D. A cranio-cervical malformation presenting as acute respiratory failure. *J. Emerg. Med.* 1996;14:569-572.

Ali N, Pitson D, Stradling J. Sleep disordered breathing: effects of adenotonsillectomy on behavior and psychological functioning. *Eur. J. Pediatr.* 1996;155:56-62.

Ali NJ, Pitson DJ, Stradline DJ. Snoring, sleep disturbance, and behaviour in 4-5 year olds. *Arch. Dis. Child.* 1993;68:360-366.

Ali NJ, Pitson D, Stradling JR. Natural history of snoring and related behaviour problems between the ages of 4 and 7 years. *Arch. Dis. Child.* 1994;71:74-76.

Ali NJ, Pitson DJ, Stradling JR. Sleep disordered breathing: effects of adenotonsillectomy on behaviour and psychological functioning. *Eur J Pediatr.* 1996;155(1):56-62.

Aljadeff G, Gozal D, Bailey-Wahl SL. Effects of overnight supplemental oxygen in obstructive sleep apnea in children. *Am. J. Respir. Crit. Care Med.* 1996;153:51-55.

Aljadeff G, Gozal D, Schechtman VL et al. Heart rate variability in children with obstructive sleep apnea. *Sleep.* 1997;20:151-157.

Allen CJ, Newhouse MT. Gastro-oesophageal reflux and chronic respiratory disease. *Am. Rev. Resp. Dis.* 1984;129:645-647.

Alvarez JA, Bodani J, Rajardo C et al. Sighs and their relationship to apnea in the newborn infant. *Biol. Neonate.* 1993;63:139-146.

Alvarez JE, Bodani J, Fajardo CA et al. Sighs and their relationship to apnea in the newborn infant. *Biol. Neonate.* 1993;63(3):139-146.

American Academy of Pediatrics. Task Force on Infant Sleep Position and Sudden Infant Death Syndrome. Changing concepts of sudden infant death syndrome: implications for infant sleeping environment and sleep position. *Pediatrics.* 2000;105(3 Pt 1):650-656.

American Academy of Pediatrics. Apnea, sudden infant death syndrome, and home monitoring. *Pediatrics.* 2003;111(4):914-917.

American Sleep Disorders Association. *International Classification of Sleep Disorders, Revised: Diagnostic and Coding Manual.* Rochester, MN: American Sleep Disorders Association; 1997.

American Thoracic Society. Standards and indications for cardiopulmonary sleep studies in children. *Am. J. Respir. Crit. Care Med.* 1996;153:866-878.

Anch AM, Remmers JE, Bunce H. Supraglottic airway resistance in normal subjects and patients with occlusive sleep apnea. *J. Appl. Physiol.* 1982;53:1158-1163.

Anders T, Emde R, Paremelee A. *A manual of standardized terminology, techniques and criteria for scoring of states of sleep and wakefulness in newborn infants.* Los Angeles: UCLA Brain Information Service, MINDS Nuerological Information Network; 1971.

Anders TF, Roffwarg HP. The effect of selective interruption and deprivation of sleep in the human newborn. *Dev. Psychobiol.* 1973;6:77-90.

Anders TF, Keener M. Developmental course of nighttime sleep-wake patterns in full-term and premature infants during the first year of life. *Sleep.* 1985;8:173-192.

Anders TF, Halpern LF, Hua J. Sleeping through the night: a developmental perspective. *Pediatrics.* 1992;90:554-560.

Anders TF, Eiben LA. Pediatric sleep disorders: A review of the past 10 years. *J. Am. Acad. Child Adoles. Psychiat.* 1997;36:9-20.

Andersen LI, Schmidt A, Bundgaard A. Pulmonary function and acid application in the esophagus. *Chest.* 1986;90:358-363.

Anderson GC. Pacifiers: The positive side. *MCN.* 1986;11:122-124.

Anderssen SH, Nicolaisen RB, Gabrielsen GW. Autonomic response to auditory stimulation. *Acta Paediatr.* 1993;82, Suppl. 389:913-918.

Anwar M, Marotta F, Fort MD et al. The ventilatory response to carbon dioxide in high risk infants. *Early Hum. Dev.* 1993;35:183-192.

Arad-Cohen N, Cohen A, Tirosh E. The relationship between gastroesophageal reflux and apnea in infants. *J. Pediatr.* 2000;137:321-326.

Arens R, Gozal D, Jain K. Prevalence of medium-chain acyl-CoA dehydrogenase deficiency in the sudden infant death syndrome. *J. Pediatr.* 1993;122:715-718.

Arens R, Gozal D, Omlin KJ et al. Hypoxic and hypercapnic ventilatory responses in Prader-Willi syndrome. *J. Appl. Physiol.* 1994;77:2224-2230.

Arens R, McDonough JM, Corbin AM et al. Linear dimentions of the upper airway structure during development. *Am. J. Crit. Care Med.* 2002;165(1):117-122.

Arguin AL, Swartz MK. Gastroesophageal reflux in infants: A primary care perspective. *Pediatr. Nurs.* 2004;30(1):45-51.

Ariagno RL, Guilleminault C, Korobkin R et al. "Near-miss" for sudden infant death syndrome infants: a clinical problem. *Pediatrics.* 1983;71:721-730.

Armstrong KL, O'Donnell H, McCallum R, Dadds M. Childhood sleep problems: association with prenatal factors and maternal distress/depression. *J Paediatr Child Health.* 1998;34(3):263-266.

Aschoff J. Zirkadiane Rhythmen des Menschen. *Arzneimittel. Forsch. Drug Res.* 1978;28:1850-1857.

Association ASD. EEG arousals: scoring rules and examples: a preliminary report from the Sleep Disorders Atlas Task Force of the American Sleep Disorders Association. *Sleep.* 1992;15:174-184.

Atkinson E, Vetere A, Grayson K. Sleep disruption in young children. The influence of temperament on the sleep patterns of pre-school children. *Child Care Health Dev.* 1995;21:233-246.

Bach V, Bourferrache B, Kremp O et al. Regulation of sleep and body temperature in response to exposure to cool and warm environments in neonates. *Pediatrics.* 1994;93:789-796.

Baculard A, Goutet JM, Le Moing G, et al. Syndrome astmatique et reflux gastro-oesophagien chez l'enfant de plus de trente mois. A propos de douze observations. *Ann. Pédiatr. (Paris).* 1986;33:19-26.

Badia P, Harsh J, Balkin T et al. Behavioral control of respiration in sleep and sleepiness due to signal-induced sleep framentation. *Psychophysiology.* 1985;22:517-524.

Badriul H, Vandenplas Y. Gastro-oesophageal reflux in infancy. *J Gastroenterol Hepatol.* 1999;14(1):13-19.

Bajanowski T, Brinkmann B. Thickness of the air-blood tissue barrier in infants. *Int. J. Legal. Med.* 2000;113(6):332-337.

Bandla HPR, Gozal D. Dynamic changes in EEG spectra during obstructive apnea in children. *Pediatr. Pulmonol.* 1999;29:359-365.

Bar A, Tarasiuk A, Segev Y et al. The effect of adenotonsillectomy on serum insulin-like growth factor-I and growth in children with obstructive sleep apnea syndrome. *J. Pediatr.* 1999;135:76-80.

Barbe F, Quera-Salva MA, McCann C et al. Sleep-related respiratory disturbances in patients with Duchenne muscular dystrophy. *Eur. Respir. J.* 1994;7:1403-1408.

Barbe F, Quera-Salva MA, de Lattre J et al. Long-term effects of nasal intermittent positive-pressure ventilation on pulmonary function and sleep architecture in patients with neuro-muscular diseases. *Chest.* 1996;110:1179-1183.

Barnes P, Fitzgerald G, Brown Mea. Nocturnal asthma and changes in circulating epinephrine, histamine and cortisol. *New Engl. J. Med.* 1980;303:263-267.

Barrington KJ, Finer NN, Torok-Both G. Dose-response relationship of doxapram in the therapy for refractory idiopathic apnea of prematurity. *Pediatrics.* 1987;80:22-27.

Barrington KJ, Finer N, Li D. Predischarge respiratory recordings in very low birth weight newborn infants. *J. Pediatr.* 1996;129(6):934-940.

Bartlett D. Origin and regulation of spontaneous deep breaths. *Resp. Physiol.* 1971;12:230-238.

Bartlett D. Effects of hypercapnia and hypoxia on laryngeal resistance to airflow. *Respir. Physiol.* 1979;37:293-302.

Bateman JRM, Pavia D, Clarke SW. The retention of lung secretions during the night in normal subjects. *Clin. Sci. Mol. Med.* 1978;55:523-527.

Beal S, Blundell H. Sudden infant death syndrome related to position in the cot. *Med. J. Aust.* 1978;2:217-218.

Beal SM, Blundell HK. Recurrence incidence of sudden infant death syndrome. *Arch. Dis. Child.* 1988;63:924-930.

Becker H, Faust M, Köhler U, Peter JH. Reversibility of severe cardiac arrhythmias in sleep apnea under nasal continuous positive airway pressure therapy. In: Peter JH, Podszus T, von Wichert P, eds. *Sleep related disorders and internal diseases.* Berlin, Heidelberg, New York, London, Paris, Tokyo: Springer Verlag; 1987:380-387.

Beckerman RC, Brouillette RT, Hunt CE. *Respiratory control disorders in infants and children.* Baltimore: Williams and Wilkins; 1992.

Beech DJ, Sibbons PD, Howard CV, van Velzen D. Lung development: number of terminal bronchiolar duct endings and gas exchange surface area in victims of sudden infant death syndrome. *Pediatr. Pulmonol.* 2001;31(5):339-343.

Begle RL, Badr S, Skatrud JB, al. e. Effect of lung inflation on pulmonary resistance during NREM sleep. *Am. Rev. Respir. Dis.* 1990;141:854-860.

Beltramini A, Hertzig M. Sleep and bedtime behavior in preschool-aged children. *Pediatrics.* 1983;71:153-158.

Benloucif S, Guico MJ, Reid KJ et al. Stability of melatonin and temperature as circadian phase markers and their relation to sleep times in humans. *J. Biol. Rhythms.* 2005;20:178-188.

Berger AJ, Mitchell RA, Severinghaus JW. Regulation of respiration, Part II. *N. Engl. J. Med.* 1977;297:138-143.

Bergman AB, Wiesner LA. Relationship of passive cigarette-smoking to sudden infant death syndrome. *Pediatrics.* 1976;58:665-668.

Berry M, Hargadon B, Morgan A, et al. Alveolar nitric oxide in adults with asthma: evidence of distal lung inflammation in refractory asthma. *Eur. Respir. J.* 2005;25:986-991.

Berry RB, Gleeson K. Respiratory arousal from sleep: mechanisms and significance. *Sleep.* 1997;20:654-675.

Bettelheim KA, Goldwater PN, Dwyer BW, al. e. Toxigenic E.coli associated with sudden infant death syndrome. *Scand. J. Infect. Dis.* 1990;22(4):467-476.

Beyaert C, Marchal F, Dousset B et al. Gastroesophageal reflux and acute life-threatening episodes: role of a central respiratory depression. *Biol Neonate.* 1995;68(2):87-90.

Blackwell CC, Weir DM, Busuttil A et al. Infection, inflammation, and the developmental stage of infants: A new hypothesis for aetiology of SIDS. In: Rognum TO, ed. *Sudden*

Infant Death Syndrome: New Trends in the Nineties. Oslo; Copenhagen; Stockholm; Boston: Scandinavian University Press; 1995:189-198.

Blanco CE, Dawes GS, Hanson MA, McCooke HB. The response to hypoxia of arterial chemoreceptors in fetal sheep and new-born lambs. *J. Physiol.* 1984;351:25-37.

Bledsoe SW, Hornbein TF. Central chemoreceptors and the regulation of their chemical environment. In: Hornbein TF, ed. *Regulation of breathing (part I)*. New York: Marcel Dekker; 1981:347-428.

Bohadana AB, Hannhart B, Teculescu DB. Nocturnal worsening of asthma and sleep-disordered breathing. *Asthma.* 2002;39:85-100.

Bohlman ME, Haponik EF, Smith PL. CT demonstration of pharyngeal narrowing in adult obstructive sleep apnea. *AJR.* 1983;140:543-548.

Boivin DB, Montplaisir J, Petit Dea. Effects of modafinil on symptomatology of human narcolepsy. *Clin. Neuropharmacol.* 1993;16:46-53.

Bolton DP, Taylor BJ, Campbell AJ et al. Rebreathing expired gases from bedding: a cause of cot death? *Arch. Dis. Child.* 1993;69:187-190.

Bond WC, Bohs C, Ebey J, Wolf S. Rhythmic heart rate variability (sinus arrythmia) related to stages of sleep. *Conditional Reflex.* 1973;8:98-107.

Bonora M, Shields GI, Knuth SL, et al. Selective depression by ethanol of upper airway respiratory motor activity in cats. *Am. Rev. Resp. Dis.* 1984;130:151-161.

Bonora M, St. John WM, Bledscoe TA. Differential elevation of protriptiline and depression by diazepam of upper airway respiratory motor activity. *Am. Rev. Resp. Dis.* 1985;131:41-45.

Borbély AA. A two process model of sleep regulation. *Hum. Nerobiol.* 1982;1:195-204.

Borghese IF, Minard KL, Thomas EB. Sleep rythmicity in premature infants: implication for developmental status. *Sleep.* 1995;18:523-530.

Botelho RV, Bittencourt LR, Rotta JM, Tufik S. Polysomnographic respiratory findings in patients with Arnold-Chiari type I malformation and basilar invagination, with or without syringomyelia: preliminary report of a series of cases. *Neurosurg. Rev.* 2000;23:151-155.

Boyle JT, Tuchman DN, Altschuler SM et al. Mechanisms for the association of gastroesophageal reflux and bronchospasm. *Am. Rev. Resp. Dis.* 1985;131 (Suppl):S 16-S 20.

Brady JP, McCann EM. Control of ventilation in subsequent siblings of victims of sudden infant death syndrome. *J. Pediatr.* 1985;106:212-217.

Bremner RM, Hoeft SF, Constantini M et al. Pharyngeal swallowing - the major factor in clearance of esophageal reflux episodes. *Ann. Surg.* 1993;218:364-370.

Brenning R. "Growing pains". *Acta Soc. Med. Ups.* 1960;65:185-201.

Brook H, Gibson A, Tappin D, Brown H. Case-control study of sudden infant death syndrome in Scotland. *BMJ.* 1997;314:1516-1520.

Brouillette RT, Thach BT. A neuromuscular mechanism maintaining extrathoracic airway patency. *J. Appl. Physiol.* 1979;46:801-808.

Brouillette RT, Thach BT. Control of genioglossus muscle inspiratory activity. *J. Appl. Physiol.* 1980;49:801-808.

Brouillette RT, Fernbach SK, Hunt CE. Obstructive sleep apnea in infants and children. *J. Pediatr.* 1982;100:31-40.

Brouillette RT, Hanson D, David R. A diagnostic approach to suspected obstructive sleep apnea in children. *J. Pediatr.* 1984;105:10-14.

Bruce EN, Cherniak NS. Central chemoreceptors. *J. Appl. Physiol.* 1987;62:389-402.

Burstein FD, Cohen SR, Scott PH et al. Surgical therapy for severe refractory sleep apnea in infants and children: application of the airway zone concept. *Plast. Reconstr. Surg.* 1995;96:34-41.

Busby KA, Mercier L, Pivik RT. Ontogenic variations in auditory arousal
threshold during sleep. *Psychophysiology.* 1994;31:182-188.

Butcovan D, Mihaila D, Cotrutz CE et al. New trends in "mystery SIDS" research. *Rev. Med. Chir. Soc. Med. Nat. Iasi.* 1997;101(3-4):123-128.

Calder NA, Williams BA, Smyth J et al. Absence of ventilatory responses to alternating breaths of mild hypoxia and air in infants who have had bronchopulmonary dysplasia: implications for the risk of sudden infant death. *Pediatr. Res.* 1994;35:677-681.

Cameron MH, Williams AL. Development and testing of scoring systems for predicting infants with high-risk of sudden infant death syndrome in Melbourne. *Aust. Paediatr. J.* 1986;Suppl.:37-45.

Canet E, Bureau MA. Chest wall diseases and dysfunction in children. In: Chernick V, ed. *Kendig's disorders of the respiratory tract in children.* Philadelphia: W.B. Saunders Company; 1990:648-672.

Canivet C, Jakobsson I, Hagander B. Infantile colic. Follow-up at four years of age: still more "emotional". *Acta Paediatr.* 2000;89:13-17.

Carey WB. Clinical applications of infant temperament measurements. *J. Pediatr.* 1972;81:823-828.

Carey WB. Night waking and temperament in infancy. *J. Pediatr.* 1974;84:756-758.

Carey WB. The importance of temperament-environment interactions for child health and development. In: Lewis M, Rosenblum L, eds. *The uncommon child.* New York: Plenum; 1981:31-55.

Carey WB. Temperament and increased weight gain in infants. *J. Dev. Behav. Pediatr.* 1985;6:128-131.

Carey WB. The difficult child. *Pediatr.Rev.* 1986;8:39-45.

Carroll JL, Loughlin GM. Obstructive sleep apnea syndrome in infants and children: clinical features and pathophysiology. In: Ferber R, Kryger M, eds. *Principles and practice of sleep medicine in the child.* Philadelphia: W. B. Saunders Company; 1995:163-191.

Casey PH, Bradley RH, Nelson JY, Whaley SA. The clinical assessment of a child's social and physical environment during health visits. *J. Dev. Behav. Pediatr.* 1988;9:333-339.

Cassidy SB, McKillop J, Morgan W. Sleep disorders in Prader-Willi syndrome. *Dysmorphol. Clin. Genet.* 1990;4:13-17.

Catto-Smith AG. Gastroesophageal reflux in children. *Aust. Fam. Physician.* 1998;27(6):465-469, 472-463.

Cavallo A. The pineal gland and human beings: relevance to pediatrics. *J. Pediatr.* 1993;123:843-851.

Challamel MJ. Fonction du sommeil paradoxal et ontogenese. *Neurophysiol. Clin.* 1992;22:117-132.

Charbonneau M, Marin JM, Olha AE et al. Changes in obstructive sleep apnea characteristics across the night. *Chest.* 1994;106:1695-1701.

Chervin RD, Dillon JE, Bassetti C et al. Symptoms of sleep disorders, inattention, and hyperactivity in children. *Sleep.* 1997;20:1185-1192.

Chervin RD, Archbold KH. Hyperactivity and polysomnographic findings in children evaluated for sleep-disordered breathing. *Sleep.* 2001;24:313-320.

Chipps BE, Mak H, Schuberth KC et al. Nocturnal oxygen saturation in normal and asthmatic children. *Pediatrics.* 1980;65:1157-1160.

Chiriboga CA. Fetal effects. *Neurol Clin.* 1993;11(3):707-728.

Cipolli C. Sleep, dreams and memory: an overview. *J. Sleep Res.* 1995;4:2-9.

Clement CD, Purpura DP, Mayer FE. *Sleep and the maturing nervous system.* New York: Academic Press; 1972.

Coccagna G, Mantovani M, Brignani F. Arterial pressure changes during spontaneous sleep in man. *Electroencephalography and Clinical Neurophysiology.* 1971;31:277-281.

Coffey MJ, Fitzgerald MX, McNicholas WT. Comparison of oxygen desaturation during sleep and exercise in patients with cystic fibrosis. *Chest.* 1991;100:659-662.

Coleman RM. Periodic movement in sleep (nocturnal myoclonus) and restless legs syndrome. In: Guilleminault C, ed. *Sleeping and waking disorders: indications and techniques.* Menlo Park, California: A-W; 1982:265-296.

Committee DCS. *International Classification of Sleep Disorders: Diagnostic and Coding Manual.* Rochester: American Sleep Disorders Association; 1990.

Coons S. Development of sleep and wakefulness during the first 6 months of life in normal infants. In: Guilleminault C, ed. *Sleep and its disorders in children.* New York: Raven Press; 1987:17-27.

Coote JH. Respiratory and circulatory control during sleep. *Journal of Experimental Biology.* 1982;100:223-244.

Corbo GM, Fuciarelli F, Foresi A, De-Benedetto F. Snoring in children: association with respiratory symptoms and passive smoking. *BMJ.* 1989;299:1491-1494.

Curzi-Dascalova L, Peirano P, Morel-Kahn F. Development of sleep states in normal premature and full-term newborns. *Dev. Psychobiol.* 1988;21:431-444.

Curzi-Dascalova L, Peirano P. Sleep states organization in small-for-gestational-age human neonates. *Brain Dysfunc.* 1989;2:45-54.

Curzi-Dascalova L, Figueroa JM, Eiselt M et al. Sleep state organization in premature infants of less than 35 weeks' gestational age. *Pediatr. Res.* 1993;34:624-628.

Dahl RE. The pharmacologic treatment of sleep disorders. *Psychiatr. Clin. North Am.* 1992;15:161-178.

Dahl RE. The impact of inadequate sleep on children's daytime cognitive function. *Semin. Pediatr. Neurol.* 1996;3:44-50.

Dahl RE. The development and disorders of sleep. *Advances in Pediatrics.* 1998;45:73-90.

Davenport PW, Cruz M, Stecenko AA, Kifle Y. Respiratory-related evoked potentials in children with life-threatening asthma. *Am. J. Respir. Crit. Care Med.* 2000;161:1830-1835.

Davidson Ward SL, Nickerson BG, van der Hal A et al. Absent hypoxic and hypercapneic arousal responses in children with myelomeningocele and apnea. *Pediatrics.* 1986;78:44-50.

Davidson Ward SL, Bautista DB, Keens TG. Hypoxic arousal responses in normal infants. *Pediatrics.* 1992a;89:860-864.

Davidson Ward SL, Bautista DB, Woo MS et al. Responses to hypoxia and hypercapnia in infants of substance-abusing mothers. *J. Pediatr.* 1992b;121:704-709.

Davidson Ward SL, Marcus CL. Obstructive sleep apnea in infants and young children. *J. Clin. Neurophysiol.* 1996;13:198-207.

Davies DP. Cot death in Hong Kong : a rare problem? *Lancet.* 1985;2:1346-1349.

Davies GM, Moscato J. Changes in lung mechanics following sighs in premature newborns without lung disease. *Pediatr. Pulmonol.* 1994;17:26-30.

Davis RS, Larsen GL, Grunstein MM. Respiratory response to intraesophageal acid infusion in asthmatic children during sleep. *J. Allergy Clin. Immunol.* 1983;72:393-398.

de Boissieu D, Dupont C, Barbet JP et al. Distinct features of upper gastrointestinal endoscopy in the newborn. *J Pediatr Gastroenterol Nutr.* 1994;18(3):334-338.

de Burgh Daly M, Angell James JE. Role of the arterial chemoreceptors in the control of the cardiovascular responses to breath-hold diving. In: Purves MJ, ed. *The peripheral arterial chemoreceptors.* London: Cambridge Univercity Press; 1975:384-407.

de Jonge GA, Burgmeijer RJF, Engelberts AC et al. Sleeping position for infants and cot death in the Netherlands 1985-91. *Arch. Dis. Child.* 1993;69(7):660-663.

Denenberg VH, Thoman EB. Evidence for a functional role for active (REM) sleep in infancy. *Sleep.* 1981;4:185-191.

Dieter JN, Field T, Hernandez-Reif M et al. Stable preterm infants gain more weight and sleep less after five days of massage therapy. *J Pediatr Psychol.* Sep 2003;28(6):403-411.

Dietrich W, Busto R, Valdes I, Loor Y. Effects of normothermic versus mild hyperthermic forebrain ischaemia in rats. *Stroke.* 1990;21:1318-1325.

DiLalla LF, Bishop EG. Differential maternal treatment of infant twins: effects of infant behaviors. *Behav Genet.* 1996;26(6):535-542.

DiMario FJ, Dunham B, Burleson JA et al. An evaluation of autonomic nervous system function in patients with Prader-Willi syndrome. *Pediatrics.* 2000;93:76-81.

Dobelle WH. 200 cases with a new breathing pacemaker dispel myths about diaphragm pacing. *Transact. Am. Soc. Art. Int. Org.* 1994;40:244-252.

Dodds WJ, Hogan WJ, Reid DP et al. A comparison between primary esophageal peristalsis following wet and dry swallows. *J. Appl. Physiol.* 1973;35:851-856.

Douglas N, White D, Pickett C et al. Hypercapnic ventilatory response in sleeping adults. *Am. Rev. Respir. Dis.* 1982a;126:758-762.

Douglas N, White D, Weil J et al. Hypoxic ventilatory response decreases during sleep in normal man. *Am. Rev. Respir. Dis.* 1982b;125:286-289.

Downey R, Perkin RM, MacQuarrie J. Upper airway resistance syndrome: sick, symptomatic but underrecognized. *Sleep.* 1993;16:620-623.

Dransfield DA, Spitzer AR, Fox WW. Episodic airway obstruction in premature infants. *Am J Dis Child.* 1983;137:441-443.

Drucker-Colin RR, McCaugh JL. *Neurobiology of sleep and memory*. New York: Academic Press; 1977.

Dure LS, Percy AK, Cheek WR, Laurent JP. Chiari type I malformation in children. *J. Pediatr.* 1989;115:573-576.

Duron B. Intercostal and diaphragmatic muscle endings and afferents. In: Hornbein TF, ed. *Regulation of breathing (part I)*. New York: Marcel Dekker; 1981:473-540.

Dusick AM. Medical outcomes in preterm infants. *Semin Perinatol.* 1997;21(3):164-177.

Eastwood PR, Szollosi I, Platt PR et al. Collapsibility of the upper airway during anesthesia with isoflurane. *Anesthesiology*. 2002;97:789-793.

Eckberg DL, Nerhed C, Wallin G. Respiratory modulation of muscle sympathetic and vagal cardiac outflow in man. *Journal of Physiology*. 1985;365:181-196.

Eichenwald EC, Aina A, Stark AR. Apnea frequency persists beyond term gestation in infants delivered at 24 to 28 weeks. *Pediatrics*. 1997;100:354-359.

Einspieler C, Widder J, Holzer A, Kenner T. The predictive value of behavioural risk factors for sudden infant death. *Early Hum Dev.* 1988;18:101-109.

Einspieler C. Abnormal spontaneous movements in infants with repeated sleep apnoeas. *Early Hum Dev.* 1994;36:31-48.

Eiselt M, Curzi-Dascalova L, Leffler C, Christova E. Sigh-related heart rate changes during sleep in premature and full-term newborns. *Neuropediatrics*. 1992;23(6):286-291.

Eiselt M, Schendel M, Witte H et al. Quantitative analysis of discontinuous EEG in premature and full-term newborns during quiet sleep. *Electroencephalogr. Clin. Neurophysiol.* 1997;103:528-534.

Eisenberg L. Experience, brain, and behavior: the importance of a head start. *Pediatrics*. 1999;103(5 Pt 1):1031-1035.

Ejnell H, Hedner J, Caidahl K et al. Increased sympathetic activity as possible etiology of hypertension and left ventricular hypertrophy in patients with obstructive sleep apnea syndrome. In: Peter JH, Penzel T, Podszus T, von Wichert P, eds. *Sleep and health risk*. Vol Springer Verlag. Berlin; 1991:341-347.

Ekbom KA. Growing pains and restless legs. *Acta Paediatr. Scand.* 1975;64:264-266.

Elias ME, Nicolson NA, Bora C. Sleep/wake patterns of breast-fed infants in the first 2 years of life. *Pediatrics*. 1986;27:322-329.

England SJ, Barlett D, Daubenspeck JA. Influence of human vocal cord movements on airflow and resistance during eupnea. *J. Appl. Physiol.* 1982;52:773-779.

England SJ, Stogren HAF. Influence of upper airway on breathing pattern and expiratory time constant in dog pups. *Respir. Physiol.* 1986;66:181-192.

Enlow DH. *Facial growth*. 3d ed. ed. Philadelphia: Saunders; 1990.

Fajardo B, Browning M, Fisher D, Paton J. Effect of nursery environment on state regulation in very low birth weight premature infants. *Infant Behav. Dev.* 1990;13:287-303.

Fanaroff AA, Wright LL, Stevenson DK et al. Very-low-birth-weight outcomes of the National Institute of Child Health and Human Development Neonatal Research Network, May 1991 through December 1992. *Am. J. Obstet. Gynecol.* 1995;173:1423-1431.

Faubion WA, Jr., Zein NN. Gastroesophageal reflux in infants and children. *Mayo Clin Proc.* 1998;73(2):166-173.

Ferber R. *Solve your child's sleep problems*. New York: Simon and Schuster; 1985.

Ferber R. Sleeplessness in children. In: Ferber R, Kryger MH, eds. *Principles and Practice of Sleep Medicine in the Child.* Philadelphia: WB Saunders; 1995a:79-89.

Ferber R. Circadian rhythm disorders in childhood. *Principles and Practice of Sleep Medicine in the Child.* Philadelphia: WB Saunders; 1995b:91-98.

Ferber R, Kryger M. *Principles and practice of sleep medicine in the child.* Philadelphia: WB Saunders; 1995.

Ferber R. Childhood sleep disorders. *Neurol. Clin.* 1996;14:493-511.

Ferber SG, Laudon M, Kuint J et al. Massage therapy by mothers enhances the adjustment of circadian rhythms to the nocturnal period in full-term infants. *J Dev Behav Pediatr.* Dec 2002;23(6):410-415.

Ferlauto JJ, Walker MW, Martin MS. Clinically significant gastroesophageal reflux in the at-risk premature neonate: relation to cognitive scores, days in the NICU, and total hospital charges. *J Perinatol.* 1998;18(6 Pt 1):455-459.

Fernbach SK, Brouillette RT, Riggs TW, Hunt CE. Radiologic evaluation of adenoids and tonsils in children with obstructive apnea: plain films and fluoroscopy. *Pediatr. Radiol.* 1983;13:258-265.

Ferraro NF. Craniofacial development and the airway during sleep. In: Loughlin GM, Carroll JL, Marcus CL, eds. *Sleep and breathinh in children. A developmental approach.* New York; Basel: Marcel Dekker; 2000:293-309.

Fett I, Amend G, Penzel T et al. Cyclical variation of heart rate in sleep apnea before and under nasal continuous positive airway pressure therapy. In: Peter JH, Penzel T, Podszus T, von Wichert P, eds. *Sleep and health risk.* Berlin, Heidelberg, New York, London, Paris, Tokyo, Hong Kong, Barcelona: Springer Verlag; 1991:237-245.

Field TM, Schanberg SM, Scafidi F et al. Tactile/kinesthetic stimulation effects on preterm neonates. *Pediatrics.* May 1986;77(5):654-658.

Filiano JJ, Kinney HC. A perspective on neuropathologic findings in victims of the sudden infant death syndrome: the triple-risk model. *Biol. Neonate.* 1994;65:194-197.

Finer NN, Barrington KJ, Hayes B, Hugh A. Obstructive, mixed, and central apnea in the neonate: Physiologic correlates. *J. Pediatr.* 1992;121:943-950.

Fleming AS, O'Day DH, Kraemer GW. Neurobiology of mother-infant interactions: experience and central nervous system plasticity across development and generations. *Neurosci Biobehav Rev.* 1999;23(5):673-685.

Fleming P, Goncalves AL, Levine MR, Woollard S. The development of stability of respiration in human infants: changes in ventilatory responses to spontaneous sighs. *J. Physiol.* 1984;347:1-16.

Fleming PJ, Cade D, Bryan MH, Bryan AC. Congenital central hypoventilation and sleep state. *Pediatrics.* 1980;66:425-428.

Fleming PJ, Gilbert R, Azaz Y et al. The interaction between bedding and sleeping position in the sudden infant death syndrome: a population based case-control study. *Br. Med. J.* 1990;301(7):85-89.

Fleming PJ, Levine MR, Azaz Y, et al. Interactions between thermoregulation and the control of respiration in infants: possible relationship to sudden infant death. *Acta Paediatr.* 1993;82, Suppl. 389:57-59.

Fleming PJ, Blair PS, Bacon C, et al. Environment of infants during sleep and risk of the sudden infant death syndrome: results of 1993-5 case-control study for confidential inquiry into still births and deaths in infancy. *BMJ.* 1996;313:191-195.

Flenley DC. Sleep in chronic obstructive lung disease. *Clin. Chest Med.* 1985;6:651-661.

Fletcher EC, DeBehnke RD, Lovoi MS, Gorin AG. Undiagnosed sleep apnea in patients with essential hypertension. *Annals of Internal Medicine.* 1985;103:190-195.

Fletcher EC, Miller J, Schaaf JW, Fletcher JG. Urinary catecholamines before and after tracheostomy in patients with obstructive sleep apnea and hypertension. *Sleep.* 1987;10:35-44.

Flores-Guevara R, Sternberg B, Peirano P et al. Respiratory pauses and periodic breathing assessed by cardiopneumography in normal infants and in SIDS siblings. *Neuropediatrics.* 1986;17:59-62.

Florey CD, Leech AM, Blackhall A. Infant feeding and mental and motor development at 18 months of age in first born singletons [see comments]. *Int J Epidemiol.* 1995;24 Suppl 1:S21-26.

Fonkalsrud EW, Ament ME. Gastroesophageal reflux in childhood. *Curr Probl Surg.* 1996;33(1):1-70.

Fox GPP, Matthews TG. Autonomic dysfunction at different ambient temperatures in infants at risk of sudden infant death syndrome. *Lancet.* 1989;4:1065-1067.

Francis PWJ, Muller NJ, Gurwitz D et al. Hemoglobin desaturation: its occurence during sleep in patients with cystic fibrosis. *Am. J. Dis. Child.* 1980;134:734-740.

Franco P, Groswasser J, Sottiaux M et al. Decreased cardiac responses to auditory stimulation during prone sleep. *Pediatrics.* 1995;97:174-178.

Franco P, Hainaut M, Pardou A et al. Prenatal exposure to cigarette smoke increases auditory arousal thresholds in infants. *Pediatr. Pulmonol.* 1996;22:426.

Franco P, Pardou A, Hassid S, Lurquin P et al. Auditory arousal thresholds are higher when infants sleep in the prone position. *J. Pediatr.* 1998a;132:240-243.

Franco P, Szliwowski H, Dramaix M, Kahn A. Polysomnographic study of the autonomic nervous system in potential victims of sudden infant death syndrome. *Clin. Auton. Res.* 1998b;8:243-249.

Franco P, Groswasser J, Hassid S et al. Prenatal exposure to cigarettes is associated with decreased arousal propensity in infants. *J. Pediatr.* 1999;135:34-38.

Franco P, Scaillet S, Wermembol V, Valente F et al. The influence of a pacifier on infants' arousals from sleep. *J. Pediatr.* 2000a;136:775-779.

Franco P, Szliwowski H, Dramaix M, Kahn A. Influence of ambient temperature on sleep characteristics and autonomic nervous control in healthy infants. *Sleep.* 2000b;23:401-407.

Franco P, Scaillet S, Chabanski S et al. Influence of ambient temperature on infants' arousability in healthy infants. *Sleep.* 2001;24(3):325-329.

Franco P, Lipshut W, Valente F et al. Cardiac autonomic characteristics in infants sleeping with the head covered by bedclothes. *Pediatr. Res.* 2002a;109:1112-1127.

Franco P, Lipshutz W, Valente F, Adams S et al. Decreased arousals in infants who sleep with the face covered by bedclothes. *Pediatrics.* 2002b;109:1112-1117.

Franco P, Seret N, van Hees JN et al. Cardiac autonomic changes during sleep in sleep-deprived infants. *Sleep.* 2003;26:845-848.

Franco P, Chabanski S, Scaillet S et al. Pacifier use modifies infant's cardiac autonomic controls during sleep. *Early Hum. Dev.* 2004;77:99-108.

Freed GE, Steinschneider A, Glassman M, Winn K. Sudden infant death syndrome prevention and an understanding of selected clinical issues. *Pediatr Clin North Am.* 1994;41(5):967-990.

Fried PA, Watkinson B, Dillon RF, Dulberg CS. Neonatal neurological status in a low-risk population after prenatal exposure to cigarettes, marijuana, and alcohol. *J Dev Behav Pediatr.* 1987;8(6):318-326.

Friesen CA, Streed CJ, Carney LA et al. Esophagitis and modified Bernstein tests in infants with apparent life-threatening events. *Pediatrics.* 1994;94:541-544.

Fujioka M, Young LW, Girdany BR. Radiographic evaluation of adenoidal size in children: adenoidal-nasopharyngeal ratio. *Am. J. Roentgenol.* 1979;133:401-404.

Fujita S, Conway W, Zorick F, Roth T. Surgical correction of anatomic abnormalities in obstructive sleep apnea syndrome: Uvulopalatopharyngoplasty. *Otolaryngol Head Neck Surg.* 1981;89:923-934.

Galland BC, Reeves G, Taylor BJ, Bolton DPG. Sleep position, autonomic function, and arousal. *Arch. Dis. Child. Fetal Neonatal Ed.* 1998;78:F189-F194.

Galvis AJ. Pulmonary edema complicating relief of upper airway obstruction. *Am. J. Emerg. Med.* 1987;5:294-297.

Ganong WF. *Pulmonary function. Review of medical physiology.* 16 ed. Norwalk: Appleton and Lange; 1993.

Garg M, Kurzner SI, Bautista D, Keens TG. Hypoxic arousal responses in infants with bronchopulmonary dysplasia. *Pediatrics.* 1988;82:59-63.

Gaultier C, Praud JP, Clement A et al. Respiration during sleep in children with COPD. *Chest.* 1985;87:168-173.

Gaultier C, Praud JP, Canet E et al. Paradoxical inward rib cage motion during rapid eye movement sleep in infants and young children. *J. Dev. Physiol.* 1987;9:391-397.

Gaultier CL, Reinberg A, Motohashi Y. Circadian rhythm in total pulmonary resistance of asthmatic children: effect of a beta-agonist agent. *Chronobiol. Int.* 1988;5:285-290.

Ghaem M, Armstrong KL, Trocki O et al. The sleep patterns in infants and young children with gastro-oesophageal reflux. *J. Paediatr. Child Health.* 1998;34:160-163.

Gibson WS, Jr., Cochran W. Otalgia in infants and children--a manifestation of gastroesophageal reflux. *Int J Pediatr Otorhinolaryngol.* 1994;28(2-3):213-218.

Gilbert R, Rudd P, Berry PJ et al. Combined effect of infection and heavy wrapping on risk of sudden unexplained infant death. *Arch. Dis. Child.* 1992;67:171-177.

Gilbert RE, Fleming PJ, Azaz Y, Rudd P. Signs of illness preceeding sudden unexpected death in infants. *BMJ.* 1990;300:1237-1239.

Gislason T, Benediktsdottir B. Snoring, apneic episodes, and nocturnal hypoxemia among children 6 months to 6 years old. *Chest.* 1995;107:963-966.

Glassman M, George D, Grill B. Gastroesophageal reflux in children. Clinical manifestations, diagnosis, and therapy. *Gastroenterol Clin North Am.* 1995;24(1):71-98.

Glaze DG, Rosen CL, Owens JA. Toward a practical definition of pediatric insomnia. *Curr. Ther. Res.* 2002;63(Suppl. B):B4-B17.

Glotzbach SF, Dale ME, Boeddiker M, Ariagno RL. Biological rhythmicity in normal infants during the first 3 months of life. *Pediatrics.* 1994;94:482-488.

Glotzbach SF, Edgar DM, Ariagno RL. Biological rhythmicity in preterm infants prior to discharge from neonatal intensive care. *Pediatrics.* 1995;95:231-237.

Goh DYT, Galster PMCL. Sleep architecture and respiratory disturbances in children with obstructive sleep apnea. *Am. J. Respir. Crit. Care Med.* 2000;162:682-686.

Goldberg DS, Ludwig IH. Congenital central hypoventilation syndrome: ocular findings in 37 children. *J. Pediatr. Ophthalmol. Strabismus.* 1996;33:175-180.

Goldberg JS, Enlow DH, Whitaker LA et al. Some anatomical characteristics in several craniofacial syndromes. *J. Oral Surg.* 1981;39:489-498.

Goldstein MF, Veza BA, Dunsky EH et al. Comparisons of peak diurnal expiratory flow variation, postbronchodilator FEV(1) responses, and methacholine inhalation challenges in the evaluation of suspected asthma. *Chest.* 2001;119:1001-1010.

Goldstein NA, Post C, Rosenfeld RM, Campbell TF. Impact of tonsillectomy and adenoidectomy on child behavior. *Arch. Otolaryngol. Head Neck Surg.* 2000;126:494-498.

Gonzales C, Almara L, Obeso A et al. Oxygen and acid chemoreception in the carotid body receptors. *Trends Neurosci.* 1992;15:146-153.

Gonzalez de Dios J, Garcia-Alix A, Cabanas F et al. Cerebellar hypoplasia in the newborn: association with respiratory control disorders and mental retardation. *Rev. Neurol.* 1995;23(123):1041-1046.

Gordon N. The therapeutics of melatonin: A paediatric perspective. *Brain Dev.* 2000;22:213-217.

Gozal D, Marcus CL, Shoseyov D, Keens TG. Peripheral chemoreceptor function in children with the congenital central hypoventilation syndrome. *J. Appl. Physiol.* 1993;74:379-387.

Gozal D, Arens R, Omlin KJ et al. Absent peripheral chemosensitivity in Prader-Willi syndrome. *J. Appl. Physiol.* 1994;77:2231-2236.

Gozal D, Arens R, Omlin KJ et al. Ventilatory response to consecutive short hypercapnic challenges in children with obstructive sleep apnea syndrome. *J. Appl. Physiol.* 1995;79:1608-1614.

Gozal D, Marcus CL, Ward SL, Keens TG. Ventilatory responses to passive leg motion in children with congenital central hypoventilation syndrome. *Am. J. Respir. Crit. Care Med.* 1996;153:761-768.

Grantham-McGregor SM, Lira PI, Ashworth A et al. The development of low birth weight term infants and the effects of the environment in northeast Brazil. *J Pediatr.* 1998;132(4):661-666.

Greisen G, Hellström-Vestas L, Lou H et al. Sleep-waking shifts and cerebral blood flow in stable pre-term infants. *Pediatr. Res.* 1985;19:1156-1159.

Griefahn B, Brode P, Blaszkewicz M, Remer T. Melatonin production during childhood and adolescence: a longitudinal study on the excretion of urinary 6-hydroxymelatonin sulfate. *J. Pineal Res.* 2002;34:26-31.

Grögaard JB. Apnea monitors. *Acta Paediatr.* 1993;82(Suppl 389):111-113.

Groswasser J, Simon T, Scaillet S et al. Reduced arousals following obstructive apnoeas in infants sleeping prone. *Pediatr. Res.* 2001;49:402-406.

Guilleminault C, Wilson RA, Dement WC. A study on cataplexy. *Arch. Neurol.* 1974;31:255-261.

Guilleminault C, Dement WC. Sleep apnea syndromes and related sleep disorders. In: Williams RL, Karacan I, eds. *Sleep disorders: diagnosis and treatment.* New York: Willey; 1978:9-28.

Guilleminault C, Ariagno R, Korobkin R et al. Mixed and obstructive sleep apnea and near miss for sudden infant death syndrome: 2. Comparison of near miss and normal control infants by age. *Pediatrics.* 1979;64:882-891.

Guilleminault C, McQuitty J, Ariagno RL et al. Congenital central alveolar hypoventilation syndrome in six infants. *Pediatrics.* 1982a;70:684-694.

Guilleminault C, Winke R, Korobkin R, Simmons B. Children and nocturnal snoring: Evaluation of the effects of sleep related respiratory resistive and daytime functioning. *Eur. J. Pediatr.* 1982b;139:165-171.

Guilleminault C, Connolly SJ, Winkle RA. Cardiac arrhythmia and conduction disturbances during sleep in 400 patients with sleep apnea syndrome. *Am J Cardiol.* 1983;52:490-494.

Guilleminault C, Connolly S, Winkle R et al. Cyclical variation of the heart rate in sleep apnea syndrome: mechanisms and usefulness of 24h electrocardiography as a screening technique. *Lancet.* 1984a;1(8369):126-131.

Guilleminault C, Motta J, Gillis AM. Sinus arrest during REM sleep in young adults. *N. Engl. J. Med.* 1984b;311:1006-1010.

Guilleminault C. *Sleep and its disorders in children.* New York: Raven Press; 1987.

Guilleminault C, Partinen M, Praud JP et al. Morphometric facial changes and obstructive sleep apnea in adolescents. *J. Pediatr.* 1989;114:997-999.

Guilleminault C, Stoohs R, Clerk A et al. From obstructive sleep apnea syndrome to upper airway resistance syndrome: consistency of daytime sleepiness. *Sleep.* 1992;15(Suppl. 6):S13-S16.

Guilleminault C, Stoohs R, Clerk A et al. A cause of daytime sleepiness: the upper airway resistance syndrome. *Chest.* 1993;104:781-787.

Guilleminault C, Pelayo R, Clerk A et al. Home nasal continuous positive airway pressure in infants with sleep-disordered breathing. *J. Pediatr.* 1995;127:905-912.

Guilleminault C, Pelayo R. Narcolepsy in prepubertal children. *Ann. Neurol.* 1998;43:135-142.

Guntheroth WG, Lohmann R, Spiers PS. Risk of sudden infant death syndrome in subsequent siblings. *J. Pediatr.* 1990;116:520-524.

Guntheroth WG. *Crib Death. The sudden infant death syndrome.* 3rd ed. Armonk, NY: Futura Publishing; 1995.

Guyton AC, Hall JE. Regulation of respiration. In: Guyton AC, Hall JE, eds. *Textbook of medical physiology.* Philadelphia: WB Saunders; 1996:525-535.

Gyulay S, Gould D, Sawyer B et al. Evaluation of a microprocessor-based portable home monitoring system to maesure breathing during sleep. *Sleep.* 1986;10:130-142.

Haddad GG, Mazza NM, Defendini R et al. Congenital failure of automatic control of ventilation, gastrointestinal motility and heart rate. *Medicine (Baltimore)*. 1978;57:517-526.

Haddad GG, Walsh EM, Leistner HL et al. Abnormal maturation of sleep states in infants with aborted sudden infant death syndrome. *Pediatr. Res.* 1981;15:1055-1057.

Haddad GG. Control of bereathing in children. In: Edelman NH, Santiago TV, eds. *Breathing disorders of sleep*. New York: Churchill Livingstone; 1986:57-80.

Haggart M, Giblin MJ. Passive smoking and colic-like behaviour in babies. *Health Visit.* 1988;61(3):81-82.

Haglund B, Cnattingius S. Cigarette smoking as a risk factor for sudden infant death syndrome: A population-based study. *Am. J. Publ. Health.* 1990;80:29-32.

Halasz P, Rajna P, Pal I et al. K-complexes and micro-arousals as functions of sleep process. In: Koella WP, Levin P, eds. *Sleep*. Basel: Karger; 1976:292-294.

Halberg F, Johnson EA, Nelson W et al. Autorhythmometry-procedures for physiologic self-measurements and their analysis. *Physiol. Teacher.* 1972;1:1-12.

Hansen DE, Vandenberg B. Neurophysiological features and differential diagnosis of sleep apnea syndrome in children. *J. Clin. Child Psychol.* 1997;26:304-310.

Haponik EF, Smith PL, Bohlman ME. Computerized tomography in obstructive sleep apnea. Correlation of airway size with physiology during sleep and wakefulness. *Am Rev Respir Dis.* 1983;127:221-226.

Harding R. Function of the larynx in the fetus and newborn. *Ann. Rev. Physiol.* 1984;46:645-659.

Harmant van Rijckervorsel G, Aubert-Tulkens G, Moulin D, Lyon G. Le syndrome de Joubert. Etude clinique et anatomopathologique. *Rev. Neurol.* 1983;139:715-724.

Harper RM, Leake B, Hoppenbrouers T et al. Polygraphic studies of normal infants and infants at risk for the sudden infant death syndrome: heart rate and variability as a function of state. *Pediatr. Res.* 1978a;12:778-785.

Harper RM, Walter DO, Leake B et al. Development of sinus arrythmia during sleeping and wake states in normal infants. *Sleep.* 1978b;1:33-48.

Harper RM, Leake B, Hoffman H et al. Periodicity of sleep states is altered in infants at risk for the sudden infant death syndrome. *Science.* 1981;213:1030-1032.

Harper RM, Schechtman VL, Kluge KA. Machine classification of infant sleep state using cardiorespiratory measures. *EEG and Clinical Neurophysiology.* 1987;67:379-387.

Harper RM, Bandler R. Finding the failure mechanism in sudden infant death syndrome. *Nature Medicine.* 1998;4:157-158.

Harper RM, Woo MA, Alger JR. Visualization of sleep influences on cerebellar and brainstem cardiac and respiratory control mechanisms. *Brain Res. Bull.* 2000;53(1):125-131.

Harper RM. The cerebellum and respiratory control. *Cerebellum.* 2002;1(1):1-2.

Harris JC, Allen RP. Is excessive daytime sleepiness characteristic of Prader-Willi syndrome? The effects of weight change. *Arch. Pediatr. Adolesc. Med.* 1996;150:1288-1293.

Harris MA, Sullivan CE. Sleep pattern and suplementary oxygen requirements in infants with chronic neonatal lung disease. *Lancet.* 1995;345:831-832.

Hartmann H, Seidenberg J, Noyes JP et al. Small airway patency in infants with apparent life-threatening events. *Eur. J. Pediatr.* 1998;157:71-74.

Hauck FR, Herman SM, Donovan M et al. Sleep environment and the risk of sudden infant death syndrome in an urban population: the Chicago Infant Mortality Study. *Pediatrics.* 2003;111(5 Part 2):1207-1214.

Heine RG, Jaquiery A, Lubitz L et al. Role of gastro-oesophageal reflux in infant irritability. *Arch. Dis. Child.* 1995;73(2):121-125.

Hendricks JC, Kline LR. Differential activation within costal diaphragm during rapid-eye-movement sleep in cats. *J. Appl. Physiol.* 1991;70:1194-1200.

Herbst JJ, Minton SD, Book LS. Gastroesophageal reflux causing respiratory distress and apnea in newborn infants. *J. Pediatr.* 1979;95:763-768.

Herschkowitz N. Brain development and nutrition. *Developmental Neurobiology. Nestlé Nutrition Workshop Series.* 1989;12:297-304.

Herschkowitz N, Kagan J, Zilles K. Neurobiological bases of behavioral development in the first year. *Neuropediatrics.* 1997;28(6):296-306.

Hershenson MB, Colin AA, Wohl MEB, Stark AR. Changes in the contribution of the rib cage to tidal breathing during infancy. *Am. Rev. Respir. Dis.* 1990;141:922-925.

Hertz G, Cataletto M, Feinsilver SH, Angulo M. Sleep and breathing patterns in patients with Prader-Willi syndrome (PWS): effects of age and gender. *Sleep.* 1993;16:366-371.

Herve P, Denjean A, Jian R et al. Intraesophageal perfusion of acid increases the bronchomotor response to methacholine and to isocapnic hyperventillation in asthmatic subjects. *Am. Rev. Resp. Dis.* 1986;134:986-989.

Hesz N, Wolraich M. Vocal-cord paralysis and brainstem dysfunction in children with spina bifida. *Dev. Med. Child Neurol.* 1985;27:528-531.

Hetzel MR, Clark TJH. Comparison of normal and asthmatic circadian rhythms in peak expiratory flow rate. *Thorax.* 1980;35:732-738.

Hill NS, Redline S, Carskadon MA et al. Sleep-disordered breathing in patients with Duchenne muscular dystrophy using negative pressure ventilators. *Chest.* 1992;102:1656-1662.

Hoban TFR. Rhythmic movement disorder in children. *CNS Spectrums.* 2003;8:135-138.

Hobson JA. The cellular basis of sleep cycle control. *Adv. Sleep Res.* 1974;1:217-250.

Hodson ME, Madden BP, Steven MH et al. Non-invasive mechanical ventilation for cystic fibrosis patients—a potential bridge to transplantation. *Eur. Respir. J.* 1991;4:524-527.

Hoffman HJ, Hunter JC, Ellish NJ et al. Adverse reproductive factors and the sudden infant death syndrome. In: Harper RM, Hoffman HJ, eds. *Sudden infant death syndrome: Risk factors and basic mechanisms.* New York: PMA Publishing Corp.; 1988:153-175.

Hoffman HJ, Hillman LS. Epidemiology of the sudden infant death syndrome: Maternal, neonatal, and postneonatal risk factors. In: Hunt CE, ed. *Clinics in perinatology: Apnea and SIDS.* Vol 19. Philadelphia: WB Saunders Company; 1992:717-738.

Holl RW, Hartman ML, Veldhuis JD et al. Thirty second sampling of plasma growth hormone in man: correlation with sleep stages. *J. Clin. Endocrinol. Metab.* 1991;72:854-861.

Hoppenbrouwers T, Ugartechea JC, Combs D et al. Studies of maternal-fetal interaction during the last trimester of pregnancy: ontogenesis of the basic rest-activity cycle. *Exp. Neurol.* 1978;61:136-153.

Hoppenbrouwers T, Hodgman J, Arakawa K et al. Sleep and waking states in infancy: normative studies. *Sleep.* 1988;11:387-401.

Hoppenbrouwers T, Hodgman JE, Arakawa K, Sterman MB. Polysomnographic sleep and waking states are similar in subsequent siblings of SIDS and control infants during the first six months of life. *Sleep.* 1989;12:265-276.

Hoppenbrouwers T, Hodgman JE. Skin temperatures and respiratory rates in subsequent siblings of infants who died of sudden infant death syndrome (SIDS). *IEEE Eng. Med. Biol.* 1992;14:2616-2617.

Hoppenbrouwers T, Hodgman JE, Cabal L. Obstructive apnea, associated patterns of movement, heart rate, and oxygenation in infants at low and increased risk for SIDS. *Pediatr. Pulmonol.* 1993;15:1-12.

Horne RSC, Sly DJ, Cranage SM et al. Effects of prematurity on arousal from sleep in the newborn infant. *Pediatr. Res.* 2000;47(4):468-474.

Horne RSC, Ferens D, Watts A et al. The prone sleeping position impairs arousability in term infants. *J. Pediatr.* 2001;138:811-816.

Horne RSC, Osborne A, Vitkovic J et al. Arousal from sleep is impaired following an infection. *Early Hum. Dev.* 2002;66:89-100.

Horner RL, Innes JA, Holden HB, Guz A. Afferent pathway(s) for pharyngeal dilator reflex to negative pressure in man: a study using upper airway anaesthesia. *J. Physiol. Lond.* 1991;436:31-44.

Horner RL. Autonomic consequences of arousal from sleep: mechanisms and implications. *Sleep.* 1996;19:S193-S195.

Hornyak M, Happe S, Trenkwalder C et al. Restless legs syndrome in childhood: A consensus proposal for diagnostic criteria. *Somnologie.* 2002;6:133-137.

Houston K, Buschang PH, Iannaccone ST, Seale NS. Craniofacial morphology of spinal muscular atrophy. *Pediatr. Res.* 1994;36:265-269.

Huffman LC, Bryan YE, del Carmen R et al. Infant temperament and cardiac vagal tone: assessments at twelve weeks of age. *Child Dev.* 1998;69:624-635.

Hughes JR. Development of sleep spindles in the first year of life. *Clin. Electroencephalogr.* 1996;37:107-115.

Hugues DM, Spier S, Rivlin J et al. Gastroesophageal reflux during sleep in asthmatic patients. *J. Pediatr.* 1983;102:666-672.

Hukins CA, Hillman DR. Daytime predictors of sleep hypoventilation in Duchenne muscular dystrophy. *Am. J. Respir. Crit. Care Med.* 2000;161:166-170.

Hulcrantz E, Lofstrand-Tidestrom B, Ahlquist-Rastad J. The epidemiology of sleep related breathing disorder in children. *Int. J. Pediatr. Otolaryngol.* 1995;32 (suppl):S63-S66.

Hunt CE, Brouillette RT, Hanson D. Theophylline improves pneumogram abnormalities in infants at risk of sudden infant death syndrome. *J. Pediatr.* 1983;103:969-974.

Hunt CE. Sudden infant death syndrome and other causes of infant mortality. Diagnosis, mechanisms, and risk for recurrence in siblings. *Am. J. Crit. Care Med.* 2001;164(3):346-357.

Hutchinson AA, Ross KR, Russell G. The effect of posture on ventillation and lung mechanisms in preterm and light-for-date infants. *Pediatrics.* 1979;64:429-432.

Hwang JC, St.John WM, Barlett D. Afferent pathways for hypoglossal and phrenic responses to changes in upper airway pressure. *Resp. Physiol.* 1984;55:341-354.

Idiopathic congenital central hypoventilation syndrome: diagnosis and management. *Am. J. Respir. Crit. Care Med.* 1999;160:368-373.

Ihle W, Loffler W, Esser G et al. [The effect of life change events on cognitive and socio-emotional development in early childhood]. *Z. Kinder. Jugendpsychiatr.* 1992;20:77-84.

Irgens LM. Risk factors for SIDS: Do they exist? In: Rognum TO, ed. *Sudden Infant Death Syndrome: New Trends in the Nineties.* Oslo; Copenhagen; Stockholm; Boston: Scandinavian University Press; 1995:99-105.

Isono S, Remmers JE, Tanaka A et al. Anatomy of pharynx in patients with obstructive sleep apnea and in normal subjects. *J. Appl. Physiol.* 1997;82:1319-1326.

Isono S, Shimada A, Utsugi M et al. Comparison of static mechanical properties of the passive pharynx between normal children and children with sleep-disordered breathing. *Am. J. Respir. Crit. Care Med.* 1998;157:1204-1212.

Isono S. Upper aiway muscle function during sleep. In: Loughlin GM, Carroll JL, Marcus CL, eds. *Sleep and breathing in children. A developmental approach.* New York; Basel: Marcel Dekker; 2000:261-291.

Issa FG, Sullivan CE. Upper airway closing pressures in obstructive sleep apnea. *J. Appl. Physiol.* 1984;57:520-527.

Jadcherla S, Shaker R. Esophageal and upper esophageal sphincter motor function in babies. *Am. J. Med.* 2001;111(8A):64-68.

Jantz JW, Blosser CD, Fruechting LA. A motor milestone change noted with a change in sleep position. *Arch. Pediatr. Adolesc. Med.* 1997;151:565-568.

Jeans WD, Fernando DCJ, Maw AR, Leighton BC. A longitudinal study of the growth of the nasopharynx and its contents in normal children. *Br. J. Radiol.* 1981;54:117-121.

Jeffery HE, Reid I, Rahilly P, Read DJC. Gastro-oesophageal reflux in "near-miss" sudden infant death infants in active but not quiet sleep. *Sleep.* 1980;3:393-399.

Jeffery HE, Megevand A, Page M. Why the prone position is a risk factor for Sudden Infant Death Syndrome. *Pediatrics.* 1999;104(2):263-269.

Jiang F, Yan C, Wu H et al. Sleep time of children from 1 month to 5 years of age in Shanghai, an epidemiological study. *Zhonghua Yi Xue Za Zhi.* 2002;82:736-739.

Johnson M. Infant and toddler sleep: a telephone survey of parents in one community. *J. Dev. Behav. Pediatr.* 1991;12:108-114.

Johnson P, Robinson JS, Salisbury D. *The onset and control of breathing after birth: foetal and neonatal physiology.* Cambridge: Cambridge University Press; 1973.

Jones BE. Basic mechanisms of sleep-wake states. In: Kryger MH, Roth T, Dement WC, eds. *Principles and practice of sleep medicine.* Philadelphia: Saunders; 1989:121-138.

Jones RAK. A controlled trial of a regularly cycled oscillating waterbed and a non-oscillating waterbed in the prevention of apnoea in the preterm infant. *Arch. Dis. Child.* 1981;56:889-891.

Jouvet M. Does a genetic programming of the brain occur during paradoxical sleep? Cerebral correlation of conscious experience. In: Bruser PA, Rougeul-Buser A, eds. *Symposium INSERM.* Vol 6. Amsterdam: Elsevier; 1978:256.

Jouvet M. Programmation génétique itérative et sommeil paradoxal. *Confrontations psychiatriques.* 1986;27:153-181.

Jouvet M. Paradoxical sleep mechanisms. *Sleep.* 1994;17:S77-S83.

Kaada B. The sudden infant death syndrome induced by 'the fear paralysis reflex'? *Med. Hypoth.* 1987;22(4):347-357.

Kahane JC. Growth of the human prepubertal and pubertal larynx. *J. Speech Hearing Res.* 1982;25:446-455.

Kahn A, Blum D, Hennart P et al. A critical comparison of the history of sudden death infants and infants hospitalised for near-miss for SIDS. *Eur. J. Pediatr.* 1984;143(2):103-107.

Kahn A, Hasaerts D, Blum D. Phenothiazine-induced sleep apneas in normal infants. *Pediatrics.* 1985a;75:844-847.

Kahn A, Mozin MJ, Casimir G et al. Insomnia and cow's milk allergy in infants. *Pediatrics.* 1985b;76:880-884.

Kahn A, Blum D, Montauk L. Polysomnographic studies and home monitoring of siblings of SIDS victims and of infants with no family history of sudden infant death. *Eur. J. Pediatr.* 1986a;145:351-356.

Kahn A, Blum D, Muller F et al. Sudden infant death syndrome in a twin: comparison of sibling histories. *Pediatrics.* 1986b;78:146-150.

Kahn A, Picard E, Blum D. Auditory arousal thresholds of normal and near-miss SIDS infants. *Dev. Med. Child Neurol.* 1986c;28:299-302.

Kahn A, Van de Merckt C, Dramaix M et al. Transepidermal water loss during sleep in infants at risk for sudden death. *Pediatrics.* 1987;80:245-250.

Kahn A, Blum D, Rebuffat E et al. Polysomnographic studies of infants who subsequently died of sudden infant death syndrome. *Pediatrics.* 1988;82:721-727.

Kahn A, Mozin M, Rebuffat E et al. Milk intolerance in children with persistent sleeplessness: a prospective double-blind crossover evaluation. *Pediatrics.* 1989;84:595-603.

Kahn A, Rebuffat E, Sottiaux M. Prevention of airway obstructions during sleep in infants with breath-holding spells by means of oral belladonna. A prospective double-blind crossover evaluation. *Sleep.* 1991a;14:432-438.

Kahn A, Rebuffat E, Sottiaux M et al. Arousals induced by proximal esophageal reflux in infants. *Sleep.* 1991b;14:39-42.

Kahn A, Groswasser J, Rebuffat E. Sleep and cardiorespiratory characteristics of infants victims of sudden death: A prospective case-control study. *Sleep.* 1992a;15:287-292.

Kahn A, Rebuffat E, Sottiaux M. Clinical problems in relation to apparent life threatening events in infants. *Sudden Infant Death. Annales Nestlé.* 1992b;50:74-80.

Kahn A, Rebuffat E, Sottiaux M. Effects of body movement restraint on cardiac response to auditory stimulation in sleeping infants. *Acta Paediatr.* 1992c;81(12):956-961.

Kahn A, Rebuffat E, Sottiaux M et al. Lack of temporal relation between acid reflux in the proximal oesophagus and cardiorespiratory events in sleeping infants. *Eur. J. Pediatr.* 1992d;151:208-121.

Kahn A, Groswasser J, Sottiaux M et al. Prone and supine body position and sleep characteristics in infants. *Pediatrics.* 1993a;91:1112-1115.

Kahn A, Groswasser J, Sottiaux M et al. Clinical problems in relation to apparent life-threatening events in infants. *Acta Paediatr. Suppl.* 1993b;389:107-110.

Kahn A, Groswasser J, Sottiaux M et al. Prenatal exposure to cigarettes in infants with obstructive sleep apneas. *Pediatrics.* 1994a;93(5):778-783.

Kahn A, Groswasser J, Sottiaux M et al. Mechanisms of obstructive sleep apneas in infants. *Biol. Neonate.* 1994b;65:235-239.

Kahn A, Groswasser J, Kelmanson I. Risk factors for SIDS: Risk factors for ALTE? From epidemiology to physiology. In: Rognum TO, ed. *Sudden Infant Death Syndrome: New Trends in the Nineties*. Oslo: Scandinavian University Press; 1995:132-137.

Kahn A, Groswasser J, Franco P et al. Factors influencing the determination of arousal thresholds in infants - a review. *Sleep Med.* 2000;1:273-278.

Kahn A. European Society for the Study and Prevention of Infant Death. Recommended clinical evaluation of infants with an Apparent Life-Threatening Event (ALTE). Consensus document of the European Society for the Study and Prevention of Infant Death, 2003. *Eur. J. Pediatr.* 2004;163(2):108-115.

Kales A, Kales JD, Sly RM et al. Sleep patterns of asthmatic children: all-night electroencephalographic studies. *J. Allerg.* 1970;46:301-308.

Kales A, Bixler EO, Cadieux RJ et al. Sleep apnoea in a hypertensive population. *Lancet.* 1984;2(8410):1005-1008.

Kamal I. Lung volume dependence of pharyngeal cross-sectional area by acoustic pharyngometry. *Otolaryngol. Head Neck Surg.* 2002;126:164-171.

Karacan I, Wolff SM, Williams RL et al. The effects of fever on sleep and dream patterns. *Psychosomatics.* 1968;9:331-339.

Kataria S, Swanson MS, Trevathan GE. Persistence of sleep disturbances in preschool children. *J. Pediatrics.* 1987;110:642-646.

Kato I, Franco P, Groswasser J et al. Frequency of obstructive and mixed sleep apneas in 1023 infants. *Sleep.* 2000;23(4):487-492.

Kato I, Franco P, Groswasser J et al. Incomplete arousal process in infants who were victims of sudden death. *Am. J. Respir. Crit. Care Med.* 2003;168(1):1298-1303.

Katz ES, McGrath S, Marcus CL. Late-onset central hypoventilation with hypothalamic dysfunction: a distinct clinical syndrome. *Pediatr. Pulmonol.* 2000;29:62-68.

Katz-Salamon M, Jonsson B, Lagercrantz H. Blunted peripheral chemoreceptor response to hyperoxia in a group of infants with bronchopulmonary dysplasia. *Pediatr. Pulmonol.* 1995;20:101-106.

Keener MA, Zeanah CH, Anders TF. Infant temperament, sleep organization, and nighttime parental interventions. *Pediatrics.* 1988;81:762-771.

Keens TG, Davidsom Ward SL. Syndromes affecting respiratory control during sleep. In: Loughlin GM, Carroll JL, Marcus CL, eds. *Sleep and breathing in children. A developmental approach*. New York: Marcel Dekker; 2000:525-553.

Kelly DH, Shannon DC. Periodic breathing in infants with near-miss sudden infant death syndrome. *Pediatrics.* 1979;63:355-360.

Kelly DH, Shannon DC. Episodic complete airway obstruction in infants. *Pediatrics.* 1981;67:823-827.

Kelly DH, Golub H, Carley D, Shannon DC. Pneumograms in infants who subsequently died of sudden infant death syndrome. *J. Pediatr.* 1986;109:249-254.

Kelmanson IA. Epidemiological aspects of sudden infant death syndrome in Leningrad: Anthropometric findings in victims. *Int. J. Epidemiol.* 1991a;20(4):944-949.

Kelmanson IA. Circadian variation of the frequency of sudden infant death syndrome and of sudden death from life-threatening conditions in infants. *Chronobiologia.* 1991b;18(4):181-186.

Kelmanson IA. Differences in somatic and organ growth rates in infants who died of sudden infant death syndrome. *J. Perinat. Med.* 1992;20(3):183-188.

Kelmanson IA. An assessment of microsocial environment of children diagnosed as "sudden infant death" using "PROCESS" inventory. *Eur. J. Pediatr.* 1993;152(8):686-690.

Kelmanson IA. *Sudden infant death syndrome: questions of clinics, diagnosis, epidemiology, identification of high risk, and prevention (in Russian)* [Doctor of Medical Sciences Thesis]. St. Petersburg, St. Petersburg State Pediatric Medical Academy; 1994.

Kelmanson IA. Brain stem gliosis in the victims of sudden infant death syndrome (SIDS): a sign of retarded maturation? *Zentralbl. Pathol.* 1994/95;140(6):449-452.

Kelmanson IA. *Polysomnographic study in infants (in Russian).* St. Petersburg: SPbSPMA Publisher; 1995.

Kelmanson IA. An assessment of behavioural characteristics in infants who died of sudden infant death syndrome using the Early Infancy Temperament Questionnaire. *Acta Paediatr.* 1996a;85:977-980.

Kelmanson IA. Retarded heart growth in the victims of sudden infant death: possible implications for lethal mechanisms. *Eur. J. Pediatr.* 1996b;155(6):440-444.

Kelmanson IA. *Sleep disordered breathing in children (in Russian).* St. Petersburg: Special Literature; 1997.

Kelmanson IA. Spontaneous modification of sleep position and behavioural features in 2-4-month-old infants. *Pädiatr. Grenzgeb.* 1999a;37(6):517-525.

Kelmanson IA. Use of a pacifier and behavioural features in 2-4-month-old infants. *Acta Paediatr.* 1999b;88(11):1258-1261.

Kelmanson IA. Parent-infant bed sharing and behavioural features in 2-4-month-old infants. *Early Child Dev. Care.* 1999c;149:1-9.

Kelmanson IA. Low birth weight and behavioural features in infants under one year old. *Early Child Dev. Care.* 1999d;155:61-70.

Kelmanson IA. Snoring, noisy breathing in sleep and daytime behaviour in 2-4-month-old infants. *Eur.J.Pediatr.* 2000a;109(10):734-739.

Kelmanson IA. Repetitive regurgitation and behavioural features in 2-4-month-old infants. *Pädiatr. Grenzgeb.* 2000b;39(5-6):465-476.

Kelmanson IA. Pacifier use and sleep position in 2 to 4 month old infants. *Klin. Pädiatr.* 2000c;212(5):273-276.

Kelmanson IA, Adulas EI. Low birth weight and sleep behaviour in two-month-old infants. *Somnologie.* 2002;6(4):155-160.

Kelmanson IA, Erman LV, Litvina SV. Maternal smoking during pregnancy and behavioural characteristics in 2-4-month-old infants. *Klin. Pädiatr.* 2002a;214(6):359-364.

Kelmanson IA, Erman LV, Litvina SV, Adulas EI. Quality of infant care and behavioural characteristics in 2-4-month-old infants. *Pediatrics and Related Topics.* 2002b;41(5):453-459.

Kelmanson IA, Groswasser J, Franco P, Kahn A. Sighs during sleep in future victims of sudden infant deaths. *Sleep and Hypnosis.* 2003;5(2):83-88.

Kelmanson IA. Temperament and sleep characteristics in two-month-old infants. *Sleep and Hypnosis.* 2004;6(2):78-84.

Kelmanson IA, Adulas EI. Environmental characteristics and sleep in two-month-old infants. *Klin. Padiatr.* 2004;214:259-263.

Kennaway DJ, Stamp GE, Goble FC. Development of melatonin production in infants and the impact of prematurity. *J. Clin. Endocrinol. Metab.* 1992;75:367-369.

Kerr SL, Kohrman MH. Polysomnographic abnormalities in Duchenne muscular dystrophy. *J. Child. Neurol.* 1994;9:332-334.

Kerr WJS, Kelly J, Geddes DAM. The areas of various surfaces in the human mouth from nine years to adulthood. *J. Dent. Res.* 1991;70:1528-1530.

Khan Y, Heckmatt JZ. Obstructive apnoeas in Duchenne muscular dystrophy. *Thorax.* 1994;49:157-161.

Kikuchi Y, Okabe S, Tamura G et al. Chemosensitivity and perception of dyspnea in patients with a history of near-fatal asthma. *N. Engl. J. Med.* 1994;330:1329-1334.

Kinney HC, Filiano JJ, Harper RM. The neuropathology of the sudden infant death syndrome. *J. Neuropathol. Exper. Neurol.* 1992;51:115-126.

Kinney HC, Filiano JJ, White WF. Medullary serotonergic network deficiency in the sudden infant death syndrome: review of a 15-year study of a single dataset. *J. Neuropathol. Exp. Neurol.* 2001;60(3):228-247.

Kirk VG, Morielli A, Brouillette RT. Sleep-disordered breathing in patients with myelomeningocele: the missed diagnosis. *Dev. Med. Child. Neurol.* 1999;41:40-43.

Kirkness JP, Christenson HK, Garlick SR et al.. Decreased surface tension of upper airway mucosal lining liquid increases upper airway patency in anaesthetised rabbits. *J. Physiol.* 2003;547:603-611.

Klackenberg G. Sleep behaviour studied longitudinally. *Acta Paediatr. Scand.* 1982;71:501-506.

Kluge KA, Harper RM, Schechtman VL et al. Spectral analysis assessment of respiratory sinus arrhythmia in normal infants and infants who subsequently died of sudden infant death syndrome. *Pediatr. Res.* 1988;24(6):677-682.

Koch LE, Biedermann H, Saternus KS. High cervical stress and apnoea. *Forensic Sci. Int.* 1998;97(1):1-9.

Koch LE, Koch H, Graumann-Brunt S et al. Heart rate changes in response to mild mechanical irritation of the high cervical spinal cord region in infants. *Forensic Sci. Int.* 2002;128(3):168-176.

Koenig JS, Thach BT. The effects of mass loading on the upper airway. *J. Appl. Physiol.* 1988;64:2294-2299.

Koniak-Griffin D, Verzemnieks I. The relationship between parental ratings of child behaviors, interaction, and the home environment. *Matern Child Nurs J.* 1995;23(2):44-56.

Korner AF, Ruppel EM, Rho JM. Effects of water beds on the sleep and motility of the theophylline-treated preterm infants. *Pediatrics.* 1982;70:861-869.

Kosko JR, Derkay CS. Uvulopalatopharyngoplasty: treatment of obstructive sleep apnea in neurologically impaired pediatric patients. *Int. J. Pediatr. Otorhinolaryngol.* 1995;32:241-246.

Kotagal S, Gibbons VP, Stith JA. Sleep abnormalities in patients with severe cerebral palsy. *Dev. Med. Child Neurol.* 1994;36:304-311.

Krol RC, Knuth SL, Barlett D. Selective reduction of genioglossal muscle activity by alcohol in normal human subjects. *Am. Rev. Respir. Dis.* 1984;129:247-250.

Krous HF. The International Standirdized Autopsy Protocol for Sudden Unexpected Infant Death. In: Rognum TO, ed. *Sudden Infant Death Syndrome: New Trends in the Nineties.* Oslo; Copenhagen; Stockholm; Boston: Scandinavian University Press; 1995:81-95.

Kryger MH, Roth T, Dement WC. *Principles and practice of sleep medicine.* Philadelphia: WB Saunders; 1989.

Ktonas PY, Bes FW, Rigoard MT et al. Developmental changes in the clustering pattern of sleep rapid eye movement activity during the first year of life: a Markov-process approach. *Electroenceph. Clin. Neurophysiol.* 1990;75:136-140.

Kuhn BR, Elliott AJ. Treatment efficacy in behavioral pediatric sleep medicine. *J. Psychosom. Res.* 2003;54(6):587-597.

Kumar M. Carnitine supplementation for preterm infants with recurrent apnea. 10/01/2005. Available at: http://www.medscape.com/viewarticle/502145. Accessed April 7, 2006.

Kumar P, Hanson MA. Re-setting of the hypoxic sensitivity of aortic chemoreceptors in the newborn lamb. *J. Dev. Physiol.* 1989;11:199-206.

Kuna ST, Insalaco G, Woodson GE. Thyroarythenoid muscle activity during wakefulness and sleep in normal adults. *J. Appl. Physiol.* 1988;65:1332-1339.

Kuna ST, Smickley J. Response of genioglossus muscle activity to nasal airway occlusion in normal sleeping adults. *J. Appl. Physiol.* 1988;64:347-353.

Kuna ST, Smickley JS, Insalaco G. Posterior cricoarytenoid muscle activity during wakefulness and sleep in normal adults. *J. Appl. Physiol.* 1990;68:1746-1754.

Labanowski M, Schmidt-Nowara W, Guilleminault C. Sleep and neuro-muscular disease: frequency of sleep-disordered breathing in a neuromuscular disease clinic population. *Neurology.* 1996;47:1173-1180.

Landstra AM, Postma DS, Boezen HM, van Aalderen WM. Role of serum cortisol levels in children with asthma. *Am. J. Respir. Crit. Care Med.* 2002;165:708-712.

Lassen K, Oei TP. Effects of maternal cigarette smoking during pregnancy on long-term physical and cognitive parameters of child development. *Addict Behav.* 1998;23(5):635-653.

Lavie P, Ben-Yosef R, Rubin A. Prevalence of sleep apnea syndrome among patients with essencial hypertension. *Am. Heart J.* 1984;108:373-376.

Leach CEA, Blair PS, Fleming PJ et al. The CESDI SUDI Research Group. Epidemiology of SIDS and explained sudden infant deaths. *Pediatrics.* 1999;104(4):e43.

Lehtonen L, Korhonen T, Korvenranta H. Temperament and sleeping patterns in colicky infants during the first year of life. *J. Dev. Behav. Pediatr.* 1994;15:416-420.

Leistner HL, Haddad GG, Epstein RA et al. Heart rate and heart rate variability during sleep in aborted sudden infant death syndrome. *J. Pediatr.* 1980;97:51-55.

Levi FA, Canon C, Blum JP et al. Circadian and/or circahemidian rhythms in nine lymphocyte-related variables from peripheral blood of healthy subjects. *J. Immunol.* 1985;134:217-222.

Lewak N, Van den Berg BJ, Beckwith JB. Sudden infant death syndrome risk factors: Prospective data review. *Clin. Pediatr.* 1979;18:404-411.

Lewin DS, England SJ, Rosen RC. Neuropsychological sequelae of obstructive sleep apnea in children. *Sleep Res.* 1996;25:278.

Lewis KW, Bosque EM. Deficient hypoxia awakening response in infants of smoking mothers: possible relationship to sudden infant death syndrome. *J. Pediatr.* 1995;127:691-699.

Li DK, Daling JR. Maternal smoking, low birth weight, and ethnicity in relation to sudden infant death syndrome. *Am. J. Epidemiol.* 1991;134:958-964.

Lijowska AS, Reed NW, Mertins Chiodini BA, Thatch BT. Sequential arousal and airway-defensive behavior in axphyxial sleep environments. *J. Appl. Physiol.* 1997;83:219-228.

Lindgren AC, Hellstrom LG, Ritzen EM, Milerad J. Growth hormone treatment increases CO2 response, ventilation and central inspiratory drive in children with Prader-Willi syndrome. *Eur. J. Pediatr.* 1999;158:936-940.

Lindgren C, Lin J, Graham BS et al. Respiratory syncytial virus infection enhances response to laryngeal chemostimulation and inhibits arousal from sleep in young lambs. *Acta Paediatr.* 1996;85:789-797.

Lipinski JF, Sallee FR, Jackson C, Sethuraman G. Dopamine agonist treatment of Tourette disorder in children: results of an open-label trial pregolide. *Mov. Disord.* 1997;12:402-407.

Lodemore M, Petersen SA, Wailoo MP. Development of nighttime temperature rhythms over the first 6 months of life. *Arch. Dis. Child.* 1991;66:521-524.

Lombroso CT. Neonatal polygraphy in full-term and preterm infants: a review of normal and abnormal findings. *J. Clin. Neurophysiol.* 1985;2:105-155.

Loredo JS, Clausen JL, Anco-Israel S, Dimsdale JE. Night-to-night variability and interscorer reliability of arousal measurements. *Sleep.* 1999;22:916-920.

Louis J, Cannard C, Bastuji H, Challamel MJ. Sleep ontogenesis revisited: a longitudinal 24-h home polygraphic study on 15 normal infants during the first two years of life. *Sleep.* 1997;20:323-333.

Lozoff B, Askew GL, Wolf AW. Cosleeping and early childhood sleep problems: effects of ethnicity and socioeconomic status. *J. Dev. Behav. Pediatr.* 1996;17:9-15.

Lucas A. Programming by early nutrition in man. In: Bock GR, Whelan J, eds. *The childhood environment and adult disease.* Chichester: John Wiley and Sons; 1991: 38-55.

Lydic R, Baghdoyan HA. The neurobiology of rapid-eye-movement sleep. In: Saunders NA, Sullivan CE, eds. *Sleep and breathing.* New York: Marcel Dekker; 1994:47-77.

MacFadyen UM. Regurgitation and sudden infant death syndrome. *Acta Paediatr.* 1993;82 (Suppl 389):98-101.

Madansky D, Edelbrock C. Cosleeping in a community sample of 2- and 3-year-old children. *Pediatrics.* 1990;86:197-203.

Madsen PL, Vostrup S. Cerebral blood flow and metabolism during sleep. *Cerebrovasc. Brain Metab. Rev.* 1991;3:281-296.

Malhotra A, Huang Y, Fogel RB et al. The male predisposition to pharyngeal collapse: importance of airway length. *Am. J. Respir. Crit. Care Med.* 2002;166:1388-1395.

Malloy MH, Kleinman JC, Land GH, Schramm WF. The association of maternal smoking with age and the cause of infant death. *Am. J. Epidemiol.* 1988;128:46-55.

Mann NP, Haddow R, Stokes L et al. Effect of night and day on preterm infants in a newborn nursery: randomized trial. *BMJ.* 1986;293:1265-1267.

Manni R, Cerveri I, Ottolini A et al. Sleep related breathing patterns in patients with spinal muscular atrophy. *Ital. J. Neurol. Sci.* 1993;14:565-569.

Marcus CL, Bautista DB, Amihyia A et al. Hypercapneic arousal responses in children with congenital central hypoventilation syndrome. *Pediatrics.* 1991a;88:993-998.

Marcus CL, Jansen MT, Poulsen MK et al. Medical and psychosocial outcome of children with congenital central hypoventilation syndrome. *J. Pediatr.* 1991b;119:888-895.

Marcus CL, Carroll JL, Koerner CB et al. Determinants of growth in children with the obstructive sleep apnea syndrome. *J. Pediatr.* 1994a;125:556-562.

Marcus CL, McColley SA, Carroll JL et al. Upper airway collapsibility in children with obstructive sleep apnea syndrome. *J. Appl. Physiol.* 1994b;77:918-924.

Marcus CL, Carroll JL, Bamford O et al. Supplemental oxygen during sleep in children with sleep-disordered breathing. *Am. J. Respir. Crit. Care Med.* 1995a;152(4, Pt 1):1297-1301.

Marcus CL, Ward SL, Mallory GB et al. Use of nasal continuous positive airway pressure as treatment of childhood obstructive sleep apnea. *J. Pediatr.* 1995b;127:88-94.

Marcus CL, Curtis S, Koerner CB et al. Evaluation of pulmonary function and polysomnography in obese children and adolescents. *Pediatr. Pulmonol.* 1996;21:176-183.

Marcus CL, Greene MG, Carroll JL. Blood pressure in children with obstructive sleep apnea. *Am. J. Respir. Crit. Care Med.* 1998a;157:1098-1103.

Marcus CL, Lutz J, Carroll JL, Bamford O. Arousal and ventilatory responses during sleep in children with obstructive sleep apnea. *J. Appl. Physiol.* 1998b;84:1926-1936.

Marcus CL, Bamford O, Bamford O, Lutz J. Response to inspiratory resistive loading during sleep in normal children and children with obstructive apnea. *J. Appl. Physiol.* 1999;87:1448-1454.

Marcus CL. Sleep-disordered breathing in children. *Am. J. Respir. Crit. Care Med.* 2001;164:16-30.

Marks K, Devenyi A, Bello M et al. Thermal head wrap for infants. *J. Pediatr.* 1985;107:956-959.

Martin RJ. The role of sleep-related breathing disorders in cardiorespiratory disease: hemodynamics. In: Peter JH, Podszus T, von Wichert P, eds. *Sleep related disorders and internal diseases.* Berlin, Heidelberg: Springer Verlag; 1987:299-312.

Martin SE, Wraith PK, Dearly IJ, Douglas NJ. The effect of nonvisible sleep fragmentation on daytime function. *Am. J. Resp. Crit. Care Med.* 1997;155:1596-1601.

Mascari MJ, Gottlieb W, Rogan PK et al. The frequency of uniparental disomy in Prader-Willi syndrome: implications for molecular diagnosis. *N. Engl. J. Med.* 1992;326:1599-1607.

Masterson J, Zucker C, Schulze K. Prone and supine positioning effects on energy expenditure and behavior of low birth weight neonates. *Pediatrics.* 1987;80:689-692.

Mathew OP, Abu-Osba YK, Thach BT. Genioglossus muscle responses to upper airway pressure changes: Afferent pathways. *J. Appl. Physiol.* 1982;52:445-450.

Mathisen B, Worrall L, Masel J et al. Feeding problems in infants with gastro-oesophageal reflux disease: a controlled study. *J. Paediatr. Child Health.* 1999;35(2):163-169.

Matturri L, Biondo B, Suarez-Mier MP, Rossi L. Brain stem lesions in the sudden infant death syndrome: variability in the hypoplasia of the arcuate nucleus. *Acta Neuropathol.* 2002;104(1):12-20.

Mayock DE. Apnea. 02/24/00. Available at: http://neonatal.peds.washington.edu/mertz/Public/Apnea.htm. Accessed April 7, 2006.

McColley S, Carroll J, Curtis S et al. High prevalence of allergic sensitization in children with habitual snoring and obstructive sleep apnea. *Chest.* 1997;111:170-173.

McColley SA, April MM, Carroll JL, Loughlin GM. Respiratory compromise after adenotonsillectomy in children with obstructive sleep apnea. *Arch. Otolaryngol. Head Neck Surg.* 1992;118:940-943.

McCulloch K, Brouillette RT, Guzzetta AJ, Hunt CE. Arousal responses in near-miss sudden infant death syndrome and in normal infants. *J. Pediatr.* 1982;101:911-917.

McGinty D, Szymusiac R. Neurobiology of sleep. In: Saunders NA, Sullivan CE, eds. *Sleep and Breathing.* New York: Marcel Dekker; 1994:1-26.

McGlashan ND. Sudden infant deaths in Tasmania, 1980-1986: A seven year prospective study. *Soc. Sci. Med.* 1989;29:1015-1026.

McKay LC, Evans KC, Frackowiak RS, Corfield DR. Neural correlates of voluntary breathing in humans. *J. Appl. Physiol.* 2003;95(3):1170-1178.

McKenna J, Mosko S, Richard C et al. Experimental studies of infant-parent co-sleeping: mutual physiological and behavioral influences and their relevance to SIDS (sudden infant death syndrome). *Early Hum. Dev.* 1994;38:187-201.

McKenna JJ, Thoman E, Anders T et al. Infant-parent cosleeping in evolutionary perspective: implications for understanding infant sleep development and SIDS. *Sleep.* 1993;16:263-282.

McKenna JJ, Mosko SS, Richard CA. Bedsharing promotes breastfeeding. *Pediatrics.* 1997;100:214-219.

McLoughlin A. Sudden infant deaths in Tameside. *Health Visitor.* 1988;61:235-237.

McNamara F, Issa FG, Sullivan CE. Arousal pattern following central and obstructive breathing abnormalities in infants and children. *J. Appl. Physiol.* 1996;81:2651-2657.

McNamara F, Sullinvan CE. Obstructive sleep apnea in infants: relation to family history of sudden infant death syndrome, apparent life-threatening events, and obstructive sleep apnea. *J. Pediatr.* 2000;136:318-323.

McNicholas WT, Coffey M, Boyle T. Effects of nasal airflow on breathing during sleep in normal humans. *Am. Rev. Respir. Dis.* 1993;147:620-623.

Medoff-Cooper B, Carey WB, McDevitt SC. The Early Infancy Temperament Questionnaire. *J. Dev. Behav. Pediatr.* 1993;14:230-235.

Mellins RB, Balfour HH, Turino GM, Winters RW. Failure of automatic control of ventilation (Ondine's curse). *Medicine.* 1970;49:487-504.

Merchant JR, Worwa C, Porter S et al. Rerspiratory instability of term and near-term healthy newborn infants in care safety seats. *Pediatrics.* 2001;108:647-652.

Mickel HS, Gilles FH. Changes in glial cells during telencephalic myelinogenesis. *Brain.* 1970;93(2):337-346.

Mignot E. Genetic and familial aspects of narcolepsy. *Neurology.* 1998;50 (2 Suppl. 1):S16-S22.

Miller JC, Horvath SM. Cardiac output during human sleep. *Aviation, Space and Environmental Medicine.* 1976;47:1046-1051.

Miller MJ, Martin RJ, Carlo WA et al. Oral breathing in newborn infants. *J. Pediatr.* 1985;107:465-469.

Miller WP. Cardiac arrhythmias and conduction disturbances in the sleep apnea syndrome. Prevalence and significance. *Am. J. Med.* 1982;73:317-321.

Milner AD, Boon AW, Saunders RA et al. Upper airway obstruction and apnea in preterm babies. *Arch. Dis. Child.* 1980;55:22-25.

Mindell JA. Sleep disorders in children. *Health Psychol.* 1993;12:151-162.

Mindell JA, Durand VM. Treatment of childhood sleep disorders: Generalization across disorders and effects on family members. *J. Pediatr. Psychol.* 1993;18:731-750.

Mindell JA, Owens JA, Carskadon MA. Developmental features of sleep. *Child Adolesc. Psychiatr. Clin. North Am.* 1999;8:695-725.

Mirmiran M, Kok JH, de Kleine MJK et al. Circadian rhythms in preterm infants: a preliminary study. *Early Hum. Dev.* 1990;23:139-146.

Mitchell EA, Taylor BJ, Ford RPK. Four modifiable and other major risk factors for cot deeath: The New Zealand study. *J. Paediatr. Child Health.* 1992;28(Suppl. 1):S3-S8.

Mitchell EA, Taylor BJ, Ford RPK et al. Dummies and the sudden infant death syndrome. *Arch. Dis. Child.* 1993;68:501-504.

Mitchell EA, Thompson JMD. Parental reported apnoea, admissions to hospital and sudden infant death syndrome. *Acta Paediatr.* 2001;90:417-422.

Mitchell RA, Berger AJ. Neural regulation of respiration. In: Hornbein TF, ed. *Regulation of breathing (part I).* New York: Marcel Dekker; 1981:541-620.

Mogayzel PJ, Carroll JL, Loughlin GM et al. Sleep-disordered breathing in children with achondroplasia. *J. Pediatr.* 1988;132:667-671.

Mograss MA, Ducharme FM, Brouillette RT. Movement/arousals. Description, classification, and relationship to sleep apnea in children. *Am. J. Respir. Crit. Care Med.* 1994;150(6, Pt 1):1690-1696.

Moon RY, Patel KM, McDermott Shaefer SJ. Sudden infant death syndrome in child care settings. *Pediatrics.* 2000;106:295-300.

Morrell MJ, Arabi Y, Zahn BR et al. Effect of surfactant on pharyngeal mechanics in sleeping humans: implications for sleep apnoea. *Eur. Respir. J.* 2002;20:451-457.

Mortimore IL, Mathur R, Douglas NJ. Effects of posture, route of respiration, and negative pressure on palatal muscle activity in humans. *J. Appl. Physiol.* 1995;79:448-454.

Mortimore IL, Douglas NJ. Palatal muscle EMG response to negative pressure in awake sleep apneic and control subjects. *Am. J. Respir. Crit. Care Med.* 1997;156:867-873.

Mosko S, Richard C, McKenna J, Drummond S. Sleep and arousals in cosleeping infants and mothers. *Sleep Res.* 1995;24:24.

Moyer-Mileur LJ, Nielson DW, Pfeffer KD et al. Eliminating sleep-associated hypoxemia improves growth in infants with bronchopulmonary dysplasia. *Pediatrics.* 1996;98:779-783.

Mukhtar AL, Cowan FM, Sothers JK. Cranial blood flow and blood pressure changes during sleep in the human neonate. *Early Hum. Dev.* 1982;6:59-64.

Muller N, Gulston G, Cade D et al. Diaphragmatic muscle fatigue in the newborn. *J. Appl. Physiol.* 1979;46:688-695.

Muller NL, Francis PW, Gurwitz D et al. Mechanism of hemoglobin desaturation during rapid-eye-movement sleep in normal subjects and in patients with cystic fibrosis. *Am. Rev. Respir. Dis.* 1980;121:463-469.

Murphy J, Newcombe R, Siebert J. The epidemiology of sudden infant death syndrome. *J. Epidemiol. Comm. Health.* 1982;36:17-21.

Myer EC, Morris DL, Adams ML et al. Increased cerebrospinal fluid beta-endorphin immunoreactivity in infants with apnea and in siblings of victims of sudden infant death syndrome. *J. Pediatr.* 1987;111:660-666.

Myerberg DZ, Carpenter RG, Myerberg CF et al. Reducing postneonatal mortality in Western Virginia: a statewide intervention program targeting risk identified at and after birth. *Am. J. Public Health.* 1995;85:631-637.

Naeye RL, Ladis B, Drage JS. Sudden infant death syndrome: A prospective study. *Am. J. Dis. Child.* 1976a;130:1207-1212.

Naeye RL, Messmer J, Specht T, Merritt TA. Sudden infant death syndrome temperament before death. *J. Pediatr.* 1976b;88:511-515.

Naeye RL. Origins of the sudden infant death syndrome. In: Tildon JT, Roeder LM, Steinschneider A, eds. *Proceedings of International Conference on SIDS held in Baltimore.* New York: Academic Press; 1983:77-83.

Naeye RL, Olsson JM, Combs JW. New brain stem and bone marrow abnormalities in victims of the sudden infant death syndrome. *J. Perinatol.* 1989;9:1980-1983.

National Institute of Health. Consensus Development Conference on Infantile Apnea and Home Monitoring, Sept 29 to Oct 1, 1986. *Pediatrics.* 1986;79:292-299.

Nattie E, Kinney H. Nicotine, serotonine, and sudden infant death syndrome. *Am. J. Respir. Crit. Care Med.* 2002;166:1530-1531.

Navelet Y, Anders T, Guilleminault C. Narcolepsy in children. In: Guilleminault C, Dement WC, Passouant P, eds. *Narcolepsy.* New York: Spectrum Publications; 1976:171-177.

Navelet Y. Insomnia in the child and adolescent. *Sleep.* 1996;19(Suppl. 3):S23-S28.

Neely CA. Effects of nonnutritive sucking upon the behavioral arousal of the newborn. In: Anderson GC, Raff B, eds. *Newborn Behavior Organization: Nursing Research and Implications.* New York: March of Dimes; 1979:173-200.

Nelson EAS. Hyperthermia: A mechanism of sudden infant death. In: Rognum TO, ed. *Sudden Infant Death Syndrome: New Trends in the Nineties.* Oslo; Copenhagen; Stockholm; Boston: Scandinavian University Press; 1995:109-113.

Nelson HS. Gastroesophageal reflux and pulmonary disease. *J. Allergy Clin. Immunol.* 1984;73:547-556.

Nelson SP, Chen EH, Syniar GM, Christoffel KK. Prevalence of symptoms of gastroesophageal reflux during infancy. A pediatric practice-based survey. Pediatric Practice Research Group. *Arch. Pediatr. Adolesc. Med.* 1997;151(6):569-572.

Newman NM, Trinder JA, Phillips KA et al. Arousal deficit: mechanism of the sudden infant death syndrome? *Aust. Paediatr. J.* 1989;25:196-201.

Nickerson B. Control of breathing problems associated with Arnold-Chiari malformation. In: Beckerman RC, Brouillette RT, Hunt CE, eds. *Respiratory control disorders in infants and children.* Baltimore: Williams and Wilkins; 1992:271-276.

Nijhuis J, Martin C, Prechtl H, Bots R. The implication of fetal behavioural states for perinatal monitoring. *Europ. J. Obstetr. Gynecol. Reprod. Biol.* 1983;15:433-436.

Noble R. Concepts of neuronal control applied to the developing brainstem. In: Hanson MA, ed. *The fetal and neonatal brainstem.* Cambridge: Cambridge University Press; 1992:48-58.

Novosad C, Freudigman K, Thoman EB. Sleep patterns in newborns and temperament at eight months: a preliminary study. *J. Dev. Behav. Pediatr.* 1999;20:99-105.

Nugent JK, Lester BM, Greene SM et al. The effects of maternal alcohol consumption and cigarette smoking during pregnancy on acoustic cry analysis. *Child Dev.* 1996;67(4):1806-1815.

Obel C, Henriksen TB, Hedegaard M et al. Smoking during pregnancy and babbling abilities of the 8-month-old infant. *Paediatr Perinat Epidemiol.* 1998;12(1):37-48.

Oberlander TF, Eckstein-Grunau R, Pitfield S et al. The developmental character of cardiac responses to an acute noxious event in 4 and 8 month–old healthy infants. *Pediatr. Res.* 1999;45:519-525.

Oehmichen M, Wullen B, Zilles K, Saternus KS. Cytological investigations on the cerebellar cortex of sudden infant death victims. *Acta Neuropathol.* 1989;78(4):404-410.

Oehmichen M. Recent neuropathologic research in sudden infant death syndrome. A critical review with special consideration of the brain stem. *Forens. Sci. Progr.* 1990;4:127-139.

Okai T, Kozuma S, Shinozuka N et al. A study on the development of sleep-wakefulness cycle in the human foetus. *Early Hum. Dev.* 1992;29:391-396.

Openshaw P, Edwards S, Helms P. Changes in rib cage geometry during childhood. *Thorax.* 1984;39:624-627.

Oren J, Kelly DH, Todres ID, Shannon DC. Respiratory complications in patients with myelodysplasia and Arnold-Chiari malformation. *Am. J. Dis. Child.* 1986;140:221-224.

Oren J, Kelly DH, Shannon DC. Pneumogram recordings in infants resuscitated for apnea of infancy. *Pediatrics.* 1989;83:364-368.

Orenstein S. An overview of reflux-associated disorders in infants: Apnea, laryngospasm, and aspiration. *Am. J. Med.* 2001;111(8A):60S-63S.

Orenstein SR, Cohn JF, Shalaby TM, Kartan R. Reliability and validity of an infant gastroesophageal reflux questionnaire. *Clin. Pediatr. (Phila).* 1993;32(8):472-484.

Orenstein SR, Shalaby TM, Cohn JF. Reflux symptoms in 100 normal infants: diagnostic validity of the infant gastroesophageal reflux questionnaire. *Clin. Pediatr. (Phila).* 1996;35(12):607-614.

Orlando R. Overview of the mechanisms of gastroesophageal reflux. *Am. J. Med.* 2001;111(8A):174-177.

Orlebeke JF, Knol DL, Verhulst FC. Increase in child behavior problems resulting from maternal smoking during pregnancy. *Arch Environ Health.* 1997;52(4):317-321.

Orr WC, Stahl ML, Duke J et al. Effect of sleep state and position on the incidence of obstructive anf central apnea in infants. *Pediatrics.* 1985;75:832-835.

Oster J, Nielsen A. Growing pains: a clinical investigation of a school population. *Arch. Paediatr. Scand.* 1972;61:329-334.

Oswald I. Sleep as a restorative process: human clues. *Prog. Brain Res.* 1980;50:279-287.

Ottaviano S, Giannotti F, Cortesi F et al. Sleep characteristics in healthy children from birth to 6 years of age in the urban area of Rome. *Sleep.* 1996;19:1-3.

Ounsted M, Hendrick A. The first-born child: patterns of development. *Dev. Med. Child Neurol.* 1977;19:446-453.

Owen G, Canter R, Robinson A. Overnight oximetry in snoring and nonsnoring children. *Clin. Otolaryngol.* 1995;20:402-406.

Owens JA, Spirito A, McGuinn M. The Children's Sleep Habits Questionnaire (CSHQ): psychometric properties of a survey instrument for school aged children. *Sleep.* 2000;23(8):1043-1051.

Owens-Stively J, Frank N, Smith A et al. Child temperament, parenting discipline style, and daytime behavior in childhood sleep disorders. *J. Dev. Behav. Pediatr.* 1997;18:314-321.

Oyemade UJ, Cole OJ, Johnson AA et al. Prenatal substance abuse and pregnancy outcomes among African American women. *J Nutr.* 1994;124(6 Suppl):994S-999S.

Oyen N, Skjaerven R, Irgens LM. Population-based recurrence risk of sudden infant death syndrome compared with other infant and fetal deaths. *Am. J. Epidemiol.* 1996;144(3):300-305.

Padman R, Lawless S, Von Nessen S. Use of BiPAP by nasal mask in the treatment of respiratory insufficiency in pediatric patients: preliminary investigation. *Pediatr. Pulmonol.* 1994;17:119-123.

Panzer SE, Dodge AM, Kelly EA, Jarjour NN. Circadian variation of sputum inflammatory cells in mild asthma. *J. Allergy Clin. Immunol.* 2003;111:308-312.

Papastamelos C, Panitch HB, England SE, Allen JL. Developmental changes in chest wall compliance in infancy and early childhood. *J. Appl. Physiol.* 1995;78:179-184.

Papastamelos C, Panitch HB, Allen JL. Chest wall compliance in infants and children with neuromuscular disease. *Am. J. Respir. Crit. Care Med.* 1996;154:1045-1048.

Parslow PM, Horne RS, Ferens D et al. Temperament ratings do not predict arousability in normal infants and infants at increased risk of sudden infant death syndrome. *J. Dev. Behav. Pediatr.* 2002;23(5):365-370.

Paton JY, Swaminathan S, Sargent CW, Keens TG. Hypoxic and hypercapnic ventilatory responses in awake children with congenital central hypoventilation syndrome. *Am. Rev. Respir. Dis.* 1989;140:368-372.

Paton JY, Nanayakkara CS, Simpson H. Observations on gastro-oesophageal reflux, central apnoea and heart rate in infants. *Eur. J. Pediatr.* 1990;149:608-612.

Paton JY, Swaminathan S, Sargent CW et al. Ventilatory response to exercise in children with congenital central hypoventilation syndrome. *Am. Rev. Respir. Dis.* 1993;147:1185-1191.

Perez-Padilla R, West P, Kryger MH. Sighs during sleep in adult humans. *Sleep.* 1983;6:234-243.

Peterson DR, Sabotta EE, Daling JR. Infant mortality among subsequent siblings of infants who died of sudden infant death syndrome. *J. Pediatr.* 1986;108:911-914.

Peterson-Falzone SJ, Pruzansky S, Parris PJ, Laffer JL. Nasopharyngeal dysmorphology in the syndromes of Apert and Crouzon. *Cleft Palate J.* 1981;18(4):237-250.

Pettigrew AG, Rahilly PM. Brain stem auditory evoked responses in infants at risk of sudden infant death. *Early Hum. Dev.* 1985;11:99-111.

Peyron C, Tighe DK, van den Pol AN et al. Neurons containing hypocretin (orexin) project to multiple neuronal systems. *J. Neurosci.* 1998;18:9996-10015.

Phillipson EA, Sullivan CE. Arousal: the forgotten response to respiratory stimuli. *Am. Rev. Respir. Dis.* 1978;118:807-808.

Picchietti DL, England SJ, Walters AS et al. Periodic limb movement disorder and restless legs syndrome in children with attention-deficit hyperactivity disorder. *J. Child. Neurol.* 1998;13:588-594.

Picchietti DL, Underwood DJ, Farris WA et al. Further studies on periodic limb movement disorder and restless legs syndrome in children with attention-deficit hyperactivity disorder. *Mov. Disord.* 1999;14:1000-1007.

Pickens DL, Schefft GL, Thach BT. Pharyngeal fluid clearance and aspiration preventive mechanisms in sleeping infants. *J. Appl. Physiol.* 1989;66:1164-1171.

Poets CF. When do infants need additional inspired oxygen? A review of the current literature. *Pediatr. Pulmonol.* 1998;26:424-428.

Ponsonby AL, Dwyer T, Gibbons LE et al. Thermal environment and sudden infant death syndrome: case-control study. *Br. Med. J.* 1992;304(2):277-282.

Ponsonby AL, Dwyer T, Couper D. Factors related to infant apnoea and cyanosis: a population based study. *J. Pediatr. Child Health.* 1997;33:317-323.

Postic WP. Comparison of polysomnography and sonography for assesing regularity of respiration during sleep in adenotonsillar hypertrophy. *Laryngoscope.* 1987;97:1430-1437.

Powell N, Guilleminault C, Riley R. Mandibular advancement and an obstructive sleep apnea. *Bull. Eur. Physiopathol. Respir.* 1983;19:607-610.

Praud JP, d'Allest AM, Nedelcoux H et al. Sleep-related abdominal muscle behavior during partial or complete obstructed breathing in prepubertal children. *Pediatr. Res.* 1989;26:347-350.

Prechtl HFR, Fargel JW, Weinmann HM, Backter HH. Postures, motility and respiration of low risk preterm infants. *Dev. Med. Child Neurol.* 1979;21:3-27.

Prechtl HFR, O'Brien MJ. Behavioral states of the full-term newborn: the emergence of a concept. In: Stratton P, ed. *Psychobiology of the human newborn.* New York: Wiley; 1982:53-73.

Prechtl HFR. Assessment of fetal neurological function and development. In: Levene MI, Bennett MJ, Jonathan P, eds. *Fetal and Neonatal Neurology and Neurosergery*. Edinburgh; London: Churchill Livingston; 1988:35-40.

Putnam PE, Orenstein SR, Pang D et al. Cricopharyngeal dysfunction associated with Chiari malformations. *Pediatrics*. 1992;89:871-876.

Quattrochi JJ, McBride PT, Yates AJ. Brainstem immaturity in sudden infant death syndrome: a quantitative rapid Golgi study of dendritic spines in 95 infants. *Brain Res.* 1985;325(1):39-48.

Ramchandani P, Wiggs L, Webb V, Stores G. A systematic review of treatments for settling problems and night waking in young children. *Brit. Med. J.* 2000;320:209-213.

Ramet J. Adaptation du rythme cardiaque au cours du sommeil. In: Gaultier C, ed. *Pathologie respiratoire du sommeil du nourrisson et de l'enfant*. Paris: Vigot; 1989:49-56.

Ramet J, Dehan M, Gaultier C. Cardiac and respiratory vagal reactivity in infants. In: Gaultier C, Escourrou P, Curzi-Dascalova L, eds. *Sleep and cardiorespiratory control.* Vol 217. Paris: John Libbey Eurotext Ltd.; 1991:145-153.

Ramet J. Cardiac and respiratory reactivity to gastroesophageal reflux: experimental data in infants. *Biol. Neonate.* 1994;65:240-246.

Raschke F, Hildebrandt G. Coordination and synchronisation in the cardiovascular-respiratory system. In: Hildebrandt G, Moog R, Raschke F, eds. *Chronobiology and chronomedicine*. Frankfurt: Lang; 1987:164-171.

Red PA, Horne RSC, Cranage SM et al. Dynamic changes in arousal threshold during sleep in the human infant. *Pediatr. Res.* 1998;43:697-703.

Redline S, Tosteson T, Boucher MA, Millman RP. Measurement of sleep-related breathing disturbances in epidemiologic studies. *Chest.* 1991;100:1281-1286.

Redline S, Tishler PV, Schluchter M et al. Risk factors for sleep-disordered breathing in children: associations with obesity, race, and respiratory problems. *Am. J. Respir. Crit. Care Med.* 1999;159:1527-1532.

Ree R, Roberts JL, Thach BT. Factors influencing regional patency and configuration of the upper airway in human infants. *J. Appl. Physiol.* 1985;58:635-644.

Res K, Spence DPS, Earis JE, Calverley PMA. Arousal responses to apneic events during non-rapid-eye-movement sleep. *Am. J. Respir. Crit. Care Med.* 1995;152:1016-1021.

Reinberg A, Schuller E, Delasnerie Nea. Rhythmes circadiens et circanuels des leucocytes, protéines totales, IgG, IgA, IgM d'adultes jeunes et sains. *Nouv. Presse Méd.* 1977;6:3819-3823.

Reinberg A, Hallek M, Hellbrugge TH. Rhythmes biologiques. In: Manciaux M, Lebovici M, eds. *L'enfant et sa santé*. Paris: Doin; 1987:337-360.

Reinhardt D, Schuhmarcher P, Fox A et al. Comparison of the effects of theophylline, prednisolone and sleep withdrawal on airway obstruction and urinary cyclic AMP- cyclic GMP excretion of asthmatic children with and without nocturnal asthma. *Int. J. Clin. Pharm. Ther. Tox.* 1980;18:399-404.

Remmers JE, deGroot WJ, Sauerland EK, Anch AM. Pathogenesis of upper airway occlusion during sleep. *J. Appl. Physiol.* 1978;44:931-938.

Remmers JE, Sterling J, Thorarinsson B, Kuna ST. Nasal airway positive pressure in patients with obstructive sleep apnea: methods and feasibility. *Am. Rev. Resp. Dis.* 1984;130:1152-1155.

Reppert SM, Weaver DR, Rivkees SA, Stopa EG. Putative melatonin receptors in a human biological clock. *Science.* 1988;242:78-81.

Reust C, Blake R. Diagnostic workup before diagnosing colic. *Arch. Fam. Med.* 2000;9(3):282-283.

Richman N. A double-blind drug trial of treatment in young children with waking problems. *J. Child Psychol. Psychiatry.* 1985;26:591-598.

Riley R, Guilleminault C, Powell N, Dermon S. Mandibular osteotomy and hyoid bone advancement for obstructive sleep apnea: a case report. *Sleep.* 1984;7:79-82.

Rintahaka PJ, Hirvonen J. The epidemiology of sudden infant death syndrome in Finland in 1969-1980. *Forens. Sci. Int.* 1986;30:219-233.

Roberts JL, Reed WR, Thach BT. Pharyngeal airway stabilizing function of the sternohyoid and sternothyroid muscles of the rabbit. *J. Appl. Physiol: Resp. Environ. Exercise Physiol.* 1984;57:1790-1795.

Robertson SS. Human cyclic motility: Fetal-newborn continuities and newborn state differences. *Dev. Psychobiol.* 1987;20:425-442.

Roffwarg HP, Muzio JN, Dement WC. Ontogenetic development of human sleep-dream cycle. *Science.* 1966;152:604-619.

Rognum TO, Saugstad OD, Oyasaeter S, Olaisen B. Elevated levels of hypoxanthine in the vitreous humor indicate prolonged cerebral hypoxia in victims of sudden infant death syndrome. *Pediatrics.* 1988;82:615-618.

Rognum TO. Co-sleeping should be discouraged! *Scand. J. Forensic Med.* 2003;9:43-44.

Rosen CL, Frost JD, Harrison GM. Infant apnea: polygraphic studies and follow-up monitoring. *Pediatrics.* 1983;71:731-736.

Rosen G, Mahowald MW, Ferber R. Sleepwalking, confusional arousals, and sleep terrors in the child. In: Ferber R, Kryger MH, eds. *Principles and practice of sleep medicine in the child.* Philadelphia: W. B. Saunders Company; 1995:99-106.

Rosen GM, Muckle RP, Mahowald MW et al. Post-operative respiratory compromise in children with obstructive sleep apnea syndrome: can it be anticipated? *Pediatrics.* 1994;93:784-788.

Rosenblum LA, Andrews MW. Influences of environmental demand on maternal behavior and infant development. *Acta Paediatr Suppl.* 1994;397:57-63.

Rosenthal L, Bishop C, Helmus T et al. Auditory awakenings thresholds in sleepy and alert individuals. *Sleep.* 1996;19:290-295.

Rowley JA, Sanders CS, Zahn BR et al. Gender differences in upper airway compliance during NREM sleep: role of neck circumference. *J. Appl. Physiol.* 2002;92:2535-2541.

Ruboyianes JM, Cruz RM. Pediatric adenotonsillectomy for obstructive sleep apnea. *Ear Nose Throat J.* 1996;75:430-433.

Rudolph CD, Mazur LJ, Liptak GS et al. Guidelines for evaluation and treatment of gastroesophageal reflux in infants and children: Recommendations of the North American Society for Pediatric Gastroenterology and Nutrition. *J. Pediatr. Gastroenterol. Nutr.* 2001;32, Suppl. 2:S1 - S31.

Ruggins NR, Milner AD. Site of upper airway obstruction in infants following an acute life-threatening event. *Pediatrics.* 1993;91:595-601.

Ryan PJ, Hilton MF, Boldy DA et al. Validation of British Thoracic Society guidelines for the diagnosis of the sleep apnoea/hypopnoea syndrome: can polysomnography be avoided? *Thorax.* 1995;50:972-975.

Sadeh A, Lavie P, Scher A et al. Actigraphic home monitoring of sleep-disturbed and control infants and young children: a new method for pediatric assessment of sleep-wake patterns. *Pediatrics.* 1991(87):494-499.

Sadeh A. Sleep and melatonin in infants: a preliminary study. *Sleep.* 1997;20:185-191.

Saigal S, Watts J, Campbell D. Randomized clinical trial of an oscillating air mattress in preterm infants: effect on apnea, growth, and development. *J. Pediatr.* 1986;109:857-864.

Sakai K. Executive mechanisms of paradoxical sleep. *Arch. Ital. Biol.* 1988;126:239-257.

Salzarulo P. Sleep patterns in infants under continuous feeding from birth. *Electroencephalogr. Clin. Neurophysiol.* 1980;49:330-336.

Salzarulo P, Chevalier A. Sleep problems in children and their relationships with early disturbances of the waking-sleeping rhythms. *Sleep.* 1983;6:47-51.

Salzarulo P, Fagioli I. Postnatal development of sleep organization in man: speculations on the emergence of the "S process". *Neurophysiol. Clin.* 1992;22:107-115.

Samson-Dolfus D, Nogues B, Verdure-Poussin A, Malleville F. Electroencephalogramme de sommeil de l'enfant normal entre 5 mois et 3 ans. *Rev. Electroenceph. Clin. Neurophysiol.* 1977;7:335-345.

Samuels MP, Poets CF, Noyes JP et al. Diagnosis and management after life-threatening events in infants and young children who received cardiopulmonary resuscitation. *BMJ.* 1993;306:489-492.

Samuels MP. Apparent life-threatening events. Pathogenesis and management. In: Loughlin GM, Carroll JL, Marcus CL, eds. *Sleep and breathing in children. A developmental approach.* New York; Basel: Marcel Dekker; 2000:423-442.

Santiago TV, Edelman NH. Opioids and breathing. *J. Appl. Physiol.* 1985;59:1675-1685.

Santin RL, Porat R. Apnea of prematurity. 10/27/2004. Available at: http://www.emedecine.com/ped/topic1157.htm. Accessed April 7, 2006.

Saternus KS, Koebke J, von Tamaska L. Neck extension as a cause of SIDS. *Forens. Sci. Int.* 1986;31:167-174.

Sawaguchi T, Franco P, Kadhim H et al. The correlation between serotonergic neurons in the brainstem and sleep apnea in SIDS victims. *Pathophysiology.* 2004;10(3-4):167-172.

Sawicka EH, Branthwaite MA. Respiration during sleep in kyphoscoliosis. *Thorax.* 1987;42:801-808.

Scharf SM. Influence of sleep state an breathing on cardiovascular function. In: Saunders NA, Sullivan CE, eds. *Sleep and breathing.* New York: Dekker; 1984:221-240.

Schechtman VL, Harper RM, Wilson AJ, Southall DP. Sleep state organization in normal infants and victims of the sudden infant death syndrome. *Pediatrics.* 1992a;89:865-870.

Schechtman VL, Raetz SL, Harper RK et al. Dynamic analysis of cardiac R-R intervals in normal infants and infants who subsequently succumbed to the sudden infant death syndrome. *Pediatr. Res.* 1992b;31(6):606-612.

Schechtman VL, Harper RK, Harper RM. Distribution of slow-wave EEG activity across nonREM sleep in infants. *Sleep.* 1994;17:316-322.

Schechtman VL, Harper RK, Harper RM. Aberrant temporal patterning of slow-wave sleep in siblings of SIDS victims. *Electroencephalogr. Clin. Neurophysiol.* 1995;94:95-102.

Scheinmann P, Le Bourgeois M, De Blic J et al. Reflux gastro-oesophagien et asthme de l'enfant. *Rev. Fr. Allerg.* 1987;27:71-76.

Scher MS, Richardson G, Coble P et al. The effects of prenatal alcohol and marijiuana exposure: disturbances in neonatal sleep cycling an arousal. *Pediatr. Res.* 1988;24:101-105.

Scher MS, Steppe DA, Dahl RE et al. Comparison of EEG-sleep measures in healthy full-term and preterm infants at matched conceptional ages. *Sleep.* 1992;15:442-448.

Scher MS. Normal electrographic-polysomnographic patterns in preterm and full-term infants. *Semin. Pediatr. Neurol.* 1996;3:2-12.

Schidlow DV, Taussig LM, Knowles MR. Cystic fibrosis foundation consensus conference report on pulmonary complications of cystic fibrosis. *Pediatr. Pulmonol.* 1993;15:187-198.

Schmitz JP, Bitonti DA, Lemke RR. Hyoid myotomy and suspension for obstructive sleep apnea syndrome. *J. Oral Maxillofac. Surg.* 1996;54:1339-1345.

Scholle S, Zwacka G, Glaser S et al. Substance P, mean apnoea duration and the sudden infant death syndrome (SIDS). *Biomed. Biochim. Acta.* 1990;49(4):249-256.

Scholle S, Scholle HC, Zwacka G. Periodic leg movements and sleep disordered breathing in children. *Somnologie.* 2001;5:153-158.

Schroeder JS, Motta J, Guilleminault C. Hemodynamic studies in sleep apnea. In: Guilleminualt C, Dement WC, eds. *Sleep apnea syndromes.* New York: Liss; 1978:177-196.

Schulte-Hobein B, Schwartz-Bickenbach D, Abt S et al. Cigarette smoke exposure and development of infants throughout the first year of life: influence of passive smoking and nursing on cotinine levels in breast milk and infant's urine. *Acta Paediatr.* 1992;81(6-7):550-557.

Schulz H, Massetani R, Fagioli I, Salzarulo P. Spontaneous awakening from sleep in infants. *Electroencephalogr. Clin. Neurophysiol.* 1985;61:267-271.

Schwartz PJ, Segantini A. Cardiac innervation, neonatal electrocardiography, and SIDS. A key for a novel preventive strategy? *Ann. N.Y. Acad. Sci.* 1988;533:210-219.

Scoring of arousals in healthy term infants (between the ages of 1 and 6 months). *J. Sleep Res.* 2005;14:37-41.

Scragg RKR, Mitchell EA, Stewart AW et al. Infant room sharing and prone sleep position in sudden infant death syndrome. *Lancet.* 1996;347:7-12.

Sekar KC, Duke JC. Sleep apnea and hypoxemia in recently weaned premature infants with and without bronchopulmonary dysplasia. *Pediatr. Pulmonol.* 1991;10:112-116.

Series F, Cormier Y, Desmeules M. Influence of passive changes of lung volume on upper airways. *J. Appl. Physiol.* 1990;68:2159-2164.

Serratto M, Harris VJ, Carr I. Upper airways obstruction. *Arch. Dis. Child.* 1981;56:153-155.

Shannon DC, Marsland DW, Gould JB et al. Central hypoventilation during quiet sleep in two infants. *Pediatrics.* 1976;57:342-346.

Shapiro CM, Catterall JR, Montgomery J et al. Do asthmatics suffer bronchoconstriction during rapid eye movement sleep? *BMJ.* 1986;292:1161-1164.

Shatz A, Hiss J, Arensburg B. Basement-membrane thickening of the vocal cords in sudden infant death syndrome. *Laryngoscope.* 1991;101:484-486.

Sheldon HS, Spire JP, Levy HB. *Pediatric sleep medicine*. Philadelphia: WB Saunders; 1992.

Shepard JW. Gas exchange and hemodynamics during sleep. *Med. Clin. North Am.* 1985;69:1243-1264.

Sher MS, Richardson GA, Salemo DG et al. Sleep architecture and continuity measures of neonates with chronic lung disease. *Sleep.* 1992;15:195-201.

Sherman MS, Kaplan JM, Effgen S et al. Pulmonary dysfunction and reduced exercise capacity in patients with myelomeningocele. *J. Pediatr.* 1997;131:413-418.

Shiihara T, Shimizu T, Mitsui T et al. Isolated sleep apnea due to Chiari type I malformation and syringomyelia. *Pediatr. Neurol.* 1995;13:266-267.

Siebert JR, Haas JE. Size of the tongue in sudden infant death syndrome. *Ann. N.Y. Acad. Sci.* 1988;533:467-468.

Simonoff E, Stores G. Controlled trial of trimeprazine tartrate for night waking. *Arch. Dis. Child.* 1987;62:253-257.

Skadberg BT, Markestad T. Behavior and physiological responses during prone and supine sleep in early infancy. *Arch. Dis. Child.* 1997;76:320-324.

Slotkin TA, Lappi SE, Tayyeb MI et al. Chronic prenatal nicotine exposure sensitizes rat brain to acute postnatal nicotine challenge as assessed with ornithine decarboxilase. *Life Sci.* 1991;49:665-670.

Slotkin TA, McCook EC, Lappi SE, Seidler FJ. Altered development of basal and forkolin-stimulated adenylate cyclase activity in brain regions of rats exposed to nicotine prenatally. *Dev. Brain Res.* 1992;68:233-239.

Smith PE, Calverley PM, Edwards RH. Hypoxemia during sleep in Duchenne muscular dystrophy. *Am. Rev. Respir. Dis.* 1988;137:884-888.

Smith PE, Edwards RH, Calverley PM. Ventilation and breathing pattern during sleep in Duchenne muscular dystrophy. *Chest.* 1989;96:1346-1351.

Smolensky MH, Reinberg A, Queng JT. The chronobiology and chronopharmacology of allergy. *Ann. Allergy.* 1981;47:237-252.

Soldatos CR, Lugaresi E. Nosology and prevalence of sleep disorders. *Semin. Neurol.* 1987;7:236-242.

Somers VK, Mark AL, Zavala DC, Abboud FM. Influence of ventilation and hypocapnia on sympathetic nerve responses to hypoxia in normal humans. *J. Appl. Physiol.* 1989;67(5):2095-2100.

Sondheimer JM. Clearance of spontaneous gastroesophageal reflux in awake and sleeping infants. *Gastroenterology.* 1989;97:821-826.

Southall DP, Talbert DG. Mechanisms for abnormal apnea of possible relevance to the sudden infant death syndrome. *Ann. N. Y. Acad. Sci.* 1988;533:329-349.

Southall DP, Plunkett MCB, Banks MW et al. Covert video recordings of life threatening child abuse: lessons for child protection. *Pediatrics.* 1997;100:735-760.

Spier S, Rivlin J, Hughes D, Levison H. The effect of oxygen on sleep, blood gases and ventilation in cystic fibrosis. *Am. Rev. Respir. Dis.* 1984;129:712-718.

Spiers PS. The influence of race and gestational age on the age at maximum risk of SIDS in infancy. *Med. Hypoth.* 2000;55:51-55.

Spitzer AR, Boyle JT, Tuchman DN et al. Awake apnea associated with gastroesophageal reflux: a specific clinical syndrome. *J. Pediatr.* 1984;104:200-205.

Spitzer AR, Fox WW. Infant apnea. *Pediatr. Clin. North Am.* 1986;33:561-581.

Spurlock CW, Hinds MW, Skaggs JW, Hernandez CE. Infant death rates among the poor and nonpoor in Kentucky, 1982 to 1983. *Pediatrics.* 1987;80(2):262-269.

Standards and indications for cardiopulmonary sleep studies in children. *Am. J. Respir. Crit. Care Med.* 1996;153:866-878.

Steinschneider A. Prolonged apnea and the sudden infant death syndrome: clinical and laboratory observations. *Pediatrics.* 1972;50(4):646-653.

Steriade M, Deschênes M, Domich L, Mulle C. Abolition of spindle oscillation in thalamic neurons disconnected from nucleus reticularis thalami. *J. Neurophysiol.* 1985;54:1473-1497.

Steriade M, Buszacki G. Parallel activation of thalamic and cortical neurons by brainstem and basal forebrain cholinergic systems. In: Steriade M, Biesold D, eds. *Brain Cholinergic Systems*. Oxford: Oxford University Press; 1990:33-62.

Sterman MB, Harper RM, Havens B et al. Quantitative analysis of infant EEG development during quiet sleep. *Electroencephalogr. Clin. Neurophysiol.* 1977;43:371-385.

Stores G, Burrows A, Crawford C. Physiological sleep disturbances in children with atopic dermatitis: a case control study. *Pediatr. Dermatol.* 1998;15:264-268.

Storm H, Nylander G, Saugstad OD. The amount of brainstem gliosis in sudden infant death syndrome (SIDS) victims correlates with maternal cigarette smoking during pregnancy. *Acta Paediatr.* 1999;88(1):13-18.

Strayer A. Chiari I malformation clinical presentation and management. *J. Neurosci. Nurs.* 2001;33:90-96.

Strohl KP, O'Cain CF, Slutsky AS. Alae nasi activation and nasal resistance in healthy subjects. *J. Appl. Physiol.* 1982;52:1432-1437.

Suen JS, Arnold JE, Brooks LJ. Adenotonsillectomy for treatment of obstructive sleep apnea in children. *Arch. Otolaryngol. Head Neck Surg.* 1995;121:525-530.

Sullivan CE, Murphy E, Kozar LF, Phillipson EA. Waking and ventilatory responses to laryngeal stimulation in sleeping dogs. *J. Appl. Physiol.* 1978;45:681-689.

Super CM, Harkness S. Temperament, development and culture. In: Plomin R, Dunn J, eds. *The study of temperament: Changes, continuities and challenges*. Hillsdale: Erlbaum Associates; 1986:131-149.

Suratt PM, Dee P, Atkinson RL. Fluoroscopic and computed tomographic features of the pharyngeal airway in obstructive sleep apnea. *Am. Rev. Resp. Dis.* 1983;127:487-492.

Sutherland ER, Ellison MC, Kraft M, Martin RJ. Altered pituitary-adrenal interaction in nocturnal asthma. *Allergy Clin. Immunol.* 2003a;112:52-57.

Sutherland ER, Ellison MC, Kraft M, Martin RJ. Elevated serum melatonin is associated with the nocturnal worsening of asthma. *J. Allergy Clin. Immunol.* 2003b;112:513-517.

Sutherland ER. Nocturnal asthma: underlying mechanisms and treatment. *Curr. Allergy Asthma Rep.* 2005;5:161-167.

Suys B, De Wolf D, Hauser B et al. Bradycardia and gastroesophageal reflux in term and preterm infants: is there any relation? *J. Pediatr.* 1994;19:187-190.

Swaminathan S, Gilsanz V, Atkinson J, Keens TG. Congenital central hypoventilation syndrome associated with multiple ganglioneuromas. *Chest.* 1989a;96:423-424.

Swaminathan S, Paton JY, Davidson Ward SL et al. Abnormal control of ventilation in adolescents with myelodysplasia. *J. Pediatr.* 1989b;115:898-903.

Tabachnick E, Muller NL, Bryan AC et al. Changes in ventillation and chest wall mechanisms during sleep in normal adolescents. *J. Appl. Physiol.* 1981;51:557-564.

Tal A, Leiberman A, Margulis G, Sofer S. Ventricular dysfunction in children with obstructive sleep apnea: radionuclide assessment. *Pediatr. Pulmonol.* 1988;4:139-143.

Tangel DJ, Mezzanotte WS, White DP. Respiratory related control of palatoglossus and levator palatini muscle activity. *J. Appl. Physiol.* 1995;78:680-688.

Tayag-Kier CE, Keenan GF, Scalzi LV et al. Sleep and periodic limb movement in sleep in juvenile fiblomyalgia. *Pediatrics.* 2001;106:E70.

Teculescu DB, Caillier I, Perrin P et al. Snoring in French preschool children. *Pediatr. Pulmonol.* 1992;13:239-244.

Thach BT, Taeusch W. Sighing in newborn human infants: role of inflation-augmenting reflex. *J. Appl. Physiol.* 1976;41:502-507.

Thach BT, Menon PA, Schefft G. Effects of negative upper airway pressure on the pattern of breathing in sleeping infants. *J. Appl. Physiol.* 1989;66:1599-1605.

The NICHD Early Child Care Research Network. Early child care and self-control, compliance, and problem behavior at twenty-four and thirty-six months. *Child Dev.* 1998;69:1145-1170.

Thomas A, Chess S. *Temperament and development*. New York: Brunner/Mazel; 1977.

Thomas DA, Poole K, McArdle EK et al. The effect of sleep deprivation on sleep states, breathing events, peripheral chemoresponsiveness and arousal propensity in healthy 3 month old infants. *Eur. Respir. J.* 1996;9:932-938.

Tilkian AG, Guilleminault C, Schroeder JS et al. Sleep-induced apnea syndrome. Prevalence of cardiac arrhythmias and their reversal after tracheostomy. *Am. J. Med.* 1977;63:348-358.

Tilkian AG, Motta J, Guilleminault C. Cardiac arrhythmias in sleep apnea. In: Guilleminault C, Dement WC, eds. *Sleep apnea syndromes*. New York: Liss; 1978:197-210.

Tirosh E, Jaffe M. Apnea of infancy, seizures, and gastroesophageal reflux: an important but infrequent association. *J. Child Neurol.* 1996;11(2):98-100.

Tirosh E, Kessel A, Jaffe M et al. Outcome of idiopathic apparent life-threatening events: infant and mother perspectives. *Pediatr. Pulmonol.* 1999;28:47-52.

Tobin JM, McCloud P, Cameron DJ. Posture and gastro-oesophageal reflux: a case for left lateral positioning. *Arch. Dis. Child.* 1997;76(3):254-258.

Tomioka K, Tomioka F. Development of circadian and ultradian rhythms of premature and full-term infants. *J. Interdiscipl. Cycle Res.* 1991;22:71-80.

Tonkin S. Sudden infant death syndrome: hypothesis of causation. *Pediatrics.* 1975;55:650-661.

Trelease RB, Sieck GC, Marks JD, Harper RM. Respiratory inhibition induced by transient hypertension during sleep in unrestrained cats. *Exp. Neurol.* 1985;90:173-186.

Trelease RB, Harper RM, Arand DL, Zimmermann EG. Use of heart rate analysis for differentiation of sleep disorders. Eighth Annual Conference of IEEE. *Eng. Med. Biol. Soc.* 1986;2:1203-1206.

Trippenbach T. Pulmonary reflexes and control of breathing during development. *Biol. Neonate.* 1994;65:205-210.

Trott GE, Badura F, Wirth S et al. [Self-assessment of prescribing behavior of psychotropic drugs for children and adolescence: results of a survey of established physicians]. *Psychiatr. Prax.* 1995;22:235-239.

Tuck SA, Remmers JE. Mechanical properties of the passive pharynx in Vietnamese pot-bellied pigs. I. Statics. *J. Appl. Physiol.* 2002a;92:2229-2235.

Tuck SA, Remmers JE. Mechanical properties of the passive pharynx in Vietnamese pot-bellied pigs. II. Dynamics. *J. Appl. Physiol.* 2002b;92:2236-2244.

Tuck SJ, Monin P, Duvivier C. Effects of a rocking bed on apnoea of prematurity. *Arch. Dis. Child.* 1982;57:475-477.

Tuffnell C, Petersen S, Wailoo M. Prone sleeping infants have a reduced ability to lose heat. *Early Hum. Dev.* 1995;43:109-116.

Upton CJ, Milner AD, Stokes GM. Upper airway patency during apnoea of prematurity. *Arch. Dis. Child.* 1992;67:419-424.

Valatx JL. Sommeils et insomnies. *Pour la Science.* 1998;243:80-87.

Valdes-Dapena M. *Sudden unexpected infant death 1970 through 1975. An evolution in understanding. US Department of Health, Education, and Welfare Publication No. (HSA) 80-5255.* Washington, D.C.: US Department of Health; 1980.

Valdes-Dapena M. Sudden infant death syndrome. Morphology update for forensic pathologists - 1985. *Forens. Sci. Int.* 1986;30(2/3):177-186.

Van Cauter E, Kerkofs M, Caufriez A et al. A qualitative estimation of growth hormone secretion in normal man: reproducibility and relation to sleep and time of day. *J. Clin. Endocrinol. Metab.* 1992;74:1441-1450.

Van de Graaf WB, Gottfried SB, Mitra J et al. Respiratory function of hyoid muscles and hyoid arch. *J. Appl. Physiol.* 1984;57:197-204.

Van den Berg M. Smoking during pregnancy and post-neonatal death. *N.Z. Med. J.* 1985;98:1075-1078.

Van der Knaap MS, Valk J. MR Imaging of the various stages of normal myelination during the first year of life. *Neuroradiology.* 1990;31:459-470.

Van der Touw T, O'Neill N, Brancatisano A et al. Respiratory-related activity of soft palate muscles: augmentation by negative upper airway pressure. *J. Appl. Physiol.* 1994;76:424-432.

Van Lunteren E, Van de Graaf WB, Parker DM et al. Nasal and pharyngeal reflex responses to negative upper airway pressure. *J. Appl. Physiol.* 1984;56:746-752.

Van Lunteren E, Haxhiu MA, Cherniack NS. Relation between upper airway volume and hyoid muscle length. *J. Appl. Physiol.* 1987;63:1443-1449.

Vandenplas Y, Belli D, Benhamou PH et al. Current concepts and issues in the management of regurgitation of infants: a reappraisal. Management guidelines from a working party. *Acta Paediatr.* 1996;85(5):531-534.

Vandenplas Y, Belli DC, Dupont C et al. The relation between gastro-oesophageal reflux, sleeping-position and sudden infant death and its impact on positional therapy. *Eur. J. Pediatr.* 1997;156(2):104-106.

Veereman-Wauters G, Bochner A, Van Caillie-Bertrand M. Gastroesophageal reflux in infants with a history of near-miss sudden infant death. *J. Pediatr. Gastroenterol. Nutr.* 1991;12:319-323.

Villa MP, Dotta A, Castello D et al. Bi-level positive airway pressure (BiPAP) ventilation in an infant with central hypoventilation syndrome. *Pediatr. Pulmonol.* 1997;24:66-69.

Visser GHH, Poelmann-Weesies G, Cohen TMN, Bekedam DJ. Fetal behavior at 30-32 weeks of gestation. *Pediatr. Res.* 1987;22:655-658.

Waggener TB, Frantz ID, Cohlan BA, Stark AR. Mixed and obstructive apneas are related to ventilatory oscillations in premature infants. *J. Appl. Physiol.* 1989;66(6):2818-2826.

Wallin BG. Human sympathetic nerve activity during normal sleep and in sleep apnoea. In: Gaultier C, Escourrou P, Curzi-Dascalova L, eds. *Sleep and cardiorespiratory control.* Vol 217. Paris: John Libbey Eurotext Ltd.; 1991:73-78.

Walters AS, Hickey K, Maltzman J et al. A questionnaire study of 138 patients with restless legs syndrome: the "night walkers" survey. *Neurology.* 1996;46:92-95.

Ward SLD, Keens TG, Chan LS et al. Sudden infant death syndrome in infants evaluated by apnea programs in California. *Pediatrics.* 1986;77(4):451-458.

Waters KA, Everett FM, Bruderer JW, Sullivan CE. Obstructive sleep apnea: the use of nasal CPAP in 80 children. *Am. J. Respir. Crit. Care Med.* 1995;152:780-785.

Waters KA, Meehan B, Huang JQ et al. Neuronal apoptosis in sudden infant death syndrome. *Pediatr. Res.* 1999;45(2):166-172.

Weese-Mayer DE, Morrow AS, Brouillette RT et al. Diaphragm pacing in infants and children: a life-table analysis of implanted components. *Am. Rev. Respir. Dis.* 1989;139:974-979.

Weese-Mayer DE, Silvestri JM, Menzies LJ et al. Congenital central hypoventilation syndrome: diagnosis, management, and long-term outcome in thirty-two children. *J. Pediatr.* 1992;120:381-387.

Weinert D, Sitka U, Minors DS, Waterhouse JM. The development of circadian rhythmicity in neonates. *Early Hum. Dev.* 1994;36:117-126.

Weintraub Z, Alvaro R, Mills S et al. Short apneas and their relationship to body movements and sighs in preterm infants. *Biol. Neonate.* 1994;66:188-194.

Weissbluth M. Sleep duration and infant temperament. *J. Pediatr.* 1981;99:817-819.

Weissbluth M. Chinese-American infant temperament and sleep duration: an ethnic comparison. *J. Dev. Behav. Pediatr.* 1982a;3:99-102.

Weissbluth M. Plasma progesteron levels, infant temperament, arousals from sleep, and the sudden infant death syndrome. *Med. Hypoth.* 1982b;9:215-222.

Weissbluth M, Davis AT, Poncher J, Reiff J. Signs of airway obstruction during sleep and behavioral, developmental, and academic problems. *J. Dev. Behav. Pediatr.* 1983;4:119-121.

Weissbluth M, Davis AT, Poncher J. Night waking in 4-to-8-month-old infants. *J. Pediatr.* 1984;104:477-480.

Weissbluth M, Green OC. Plasma progesterone concentrations and infant temperament. *J. Dev. Behav. Pediatr.* 1984;5:251-253.

Weissbluth M. Sleep-temperament interactions. In: Carey WB, McDevitt SC, eds. *Clinical and educational applications of temperament research.* Amsterdam: Swets and Zeitlinger; 1989:113-116.

Weissbluth M. Naps in children: 6 months - 7 years. *Sleep.* 1995a;18:82-87.

Weissbluth M. Colic. In: Ferber R, Kryger M, eds. *Principles and practice of sleep medicine in the child.* Philadelphia: WB Saunders; 1995b:75-78.

Wells JC, Stanley M, Laidlaw AS et al. Investigation of the relationship between infant temperament and later body composition. *Int. J. Obes. Relat. Metab. Disord.* 1997;21:400-406.

Wennergren G, Milerad J, Lagercrantz H et al. The epidemiology of sudden infant death syndrome and attacks of lifelessness in Sweden. *Acta Paediatr. Scand.* 1987;76:898-906.

West JB. *Respiratory physiology - The essentials.* Baltimore; London; Los Angeles; Sydney: Williams and Wilkins; 1985.

West JB. Mechanics of breathing. In: Satterfield TS, ed. *Respiratory physiology.* Baltimore: William and Wilkins; 1990:87-89.

White DP, Lombard RM, Cadieux RJ, Zwillich CW. Pharyngeal resistance in normal humans: influence of gender, age, and obesity. *J. Appl. Physiol.* 1985;5:365-371.

Wiegand DA, Latz B, Zwillich CW et al. Geniohyoid muscle activity in normal men during wakefulness and sleep. *J. Appl. Physiol.* 1990;69:1262-1269.

Wierenga H, Brand R, Geudeke T et al. Prenatal risk factors for cot death in very preterm and small for gestational age infants. *Early Hum. Dev.* 1990;23:15-26.

Willens TE, Biederman J, Spencer T. Clonidine for sleep disturbances associated with attention deficit hyperactivity disorder. *J. Am. Acad. Child Adolesc. Psychiatry.* 1994;33:424-426.

Williams BA, Smyth J, Boon AW et al. Development of respiratory chemoreflexes in response to alterations of fractional inspired oxygen in the newborn infant. *J. Physiol.* 1991;442:81-90.

Willinger M, James LS, Catz C. Defining the sudden infant death syndrome (SIDS): Deliberations of an expert panel convened by the National Institute of Child Health and Human Development. *Paediatric Pathol.* 1991;11:677-684.

Wilson NM, Charette L, Thomson AH et al. Gastroesophageal reflux and childhood asthma: the acid test. *Thorax.* 1985;40:592-597.

Wilson SL, Thach BT, Brouillette RT, Abu-Osba YK. Upper airway patency in the human infant: Influence of airway pressure and posture. *J. Appl. Physiol. Resp. Environ. Exercise Physiol.* 1980;48:500-504.

Winkelman JW. The evoked heart rate response to periodic leg movements of sleep. *Sleep.* 1999;22:575-580.

Woo MS, Woo MA, Gozal D et al. Heart rate variability in congenital central hypoventilation syndrome. *Pediatr. Res.* 1992;31:291-296.

Woodward S, Tauber ES, Spielmann AJ, Thorpy MJ. Effects of otolithic vestibular stimulation on sleep. *Sleep.* 1990;13:533-537.

Xu F, Frazier DT. Role of the cerebellar deep nuclei in respiratory modulation. *Cerebellum.* 2002;1(1):35-40.

Yamauchi M, Nakano H, Maekawa J et al. Oxidative stress in obstructive sleep apnea. *Chest.* 2005;127:1674-1679.

Yellon RF, Goldberg H. Update on gastroesophageal reflux disease in pediatric airway disorders. *Am. J. Med.* 2001;111(8A):78S-84S.

Zuckerman B, Stevenson J, Bailey V. Sleep problems in early childhood: Continuities, predictive factors, and behavioral correlates. *Pediatrics.* 1987;80:664-671.

Zwillich C, Devlin T, While D. Bradycardia during sleep apnea: its characteristics and mechannisms. *J. Clin. Invest.* 1982;69:1286-1292.

Index

B

C

D

I

Q

R

S

T

U

V

W

Y